T0224833

PRINCIPLES OF MAGNETOSTATICS

The subject of magnetostatics—the mathematical theory that describes the forces and fields resulting from the steady flow of electrical currents—has a long history. By capturing the basic concepts, and building toward the computation of magnetic fields, this book is a self-contained discussion of the major subjects in magnetostatics.

Overviews of Maxwell's equations, the Poisson equation, and boundary value problems pave the way for dealing with fields from transverse, axial, and periodic magnetic arrangements and assemblies of permanent magnets. Examples from accelerator and beam physics give up-to-date context to the theory. Furthermore, both complex contour integration and numerical techniques (including finite difference, finite element, and integral equation methods) for calculating magnetic fields are discussed in detail with plentiful examples.

Both theoretical and practical information on carefully selected topics make this a one-stop reference for magnet designers, as well as for physics and electrical engineering students. This title, first published in 2017, has been reissued as an Open Access publication on Cambridge Core.

RICHARD C. FERNOW received his PhD at Syracuse University for work on particle physics and worked at Brookhaven National Laboratory. He contributed to the optimization of the coil design for collider magnets and made calculations of magnetic fields in solenoid channels. He is a member of the American Physical Society.

PRINCIPLES OF MAGNETOSTATICS

RICHARD C. FERNOW

Formerly Brookhaven National Laboratory

CAMBRIDGE
UNIVERSITY PRESS

Shaftesbury Road, Cambridge CB2 8EA, United Kingdom

One Liberty Plaza, 20th Floor, New York, NY 10006, USA

477 Williamstown Road, Port Melbourne, VIC 3207, Australia

314–321, 3rd Floor, Plot 3, Splendor Forum, Jasola District Centre, New Delhi – 110025, India

103 Penang Road, #05–06/07, Visioncrest Commercial, Singapore 238467

Cambridge University Press is part of Cambridge University Press & Assessment, a department of the University of Cambridge.

We share the University's mission to contribute to society through the pursuit of education, learning and research at the highest international levels of excellence.

www.cambridge.org
Information on this title: www.cambridge.org/9781009291149

DOI: 10.1017/9781009291156

First published 2017
Reissued as OA 2022

A catalogue record for this publication is available from the British Library.

ISBN 978-1-009-29114-9 Hardback
ISBN 978-1-009-29116-3 Paperback

Cambridge University Press & Assessment has no responsibility for the persistence or accuracy of URLs for external or third-party internet websites referred to in this publication and does not guarantee that any content on such websites is, or will remain, accurate or appropriate.

Contents

Preface

My career at Brookhaven National Laboratory began in 1978, when I was hired to work on a particle physics experiment at the alternating gradient synchrotron (AGS). At that time, one of the lab's major projects was a high-energy proton collider known as ISABELLE. Unfortunately, development of the superconducting magnets proposed for the project ran into serious technical difficulties. As a result, a member of the physics department, Dr. Robert Palmer, proposed a radically different design for the collider magnets. He began recruiting a small group from the physics and accelerator departments to work on the alternate magnet design. I was one of the staff members who joined his group.

Because I came from a background in particle physics, the work on the design of high-field magnets was a revelation to me. One of my main responsibilities was to work on the optimization of the dipole conductor cross-section and end designs needed to achieve the demanding field quality requirements for the accelerator. I quickly discovered that the methods needed for practical magnetic field design went far beyond my academic training in electricity and magnetism. Much of the work required frequent feedback with the engineers working on the project. Although I moved on from the magnet division after about four years, my interest in calculating magnetic fields remained with me throughout the rest of my career. A significant part of the contents of this book are based on my notes from that period.

The subject of magnetostatics has a long history. There is a vast literature, so a book of this size has to make difficult choices about which topics to include. My primary objective was to produce a self-contained discussion of the major subjects in magnetostatics with an emphasis on the computation of magnetic fields. For that reason, I have included brief treatments of most standard background material, such as the magnetostatic Maxwell's equations and potential theory. However, the choice of example topics relies heavily on my background and interests. Many of the examples come from the fields of accelerator and beam physics. I also felt it was important to expose a wider audience to a series of very insightful papers by the late Klaus Halbach of Lawrence Berkeley Laboratory. For the discussion of numerical methods, I decided to concentrate on a small number of subjects while including sufficient details to enable readers to begin writing their own computer codes, if they so desired.

The first three chapters are mostly a survey of basic material. Chapter 1 treats the theory of magnetic fields from conductors in free space, and Chapter 2 discusses fields from magnetic materials. Chapter 3 introduces the vector and scalar potentials. It includes general solutions to the Laplace equation and the solution of boundary value problems.

Chapters 4–6 discuss transverse fields in two dimensions. Chapter 4 looks at fields from line currents, current sheets, and current blocks. Field quality is introduced in terms of multipole expansions, and the effects of approximating ideal current distributions are described. Chapter 5 looks at transverse fields using complex variable methods. The powerful techniques for computing the fields of block conductors using contour integration are discussed in detail. Contour integrals are also derived for the fields from image currents and magnetized bodies. Chapter 6 looks at transverse fields that are determined by the shape of the iron. The discussion here mainly concerns the iron surfaces used in dipole and quadrupole magnets.

Chapters 7–9 discuss some other field configurations. Chapter 7 looks at axial field arrangements. This includes the fields from current loops, solenoids, and systems of coils. The solution of the solenoid field using the sheet model is treated in detail. Chapter 8 considers periodic magnetic channels. First the field from a helical conductor winding is discussed. Then inverse problems are introduced, and some of the field configurations used for magnetic wigglers and particle beam-focusing channels are examined. Chapter 9 begins with a standard treatment of the properties of permanent magnets. This is followed by a discussion of Halbach's model for rare-earth cobalt magnets and his analysis of assemblies of permanent magnets.

In Chapter 10, we relax the strict conditions of magnetostatics and allow for the case of slowly varying currents. This leads to brief discussions of some standard subjects such as Faraday's law, but also some more engineering-related topics such as eddy currents and the skin effect. This also seemed an appropriate place to include a brief discussion of magnetic field measurements using rotating coils.

Chapter 11 discusses numerical methods. No attempt is made here to survey the thousands of papers devoted to numerical solutions of magnetic field problems. Instead, three methods for solving the Poisson equation are presented with a significant amount of detail. This chapter also includes a discussion of the POISSON code, which is freely available and extremely useful for investigating

2D problems. The chapter ends by returning to the inverse problem and presenting several examples of using optimization methods.

The appendices collect some important details about mathematical techniques and special functions used in the book.

The level of the treatment of background magnetostatic topics in the book is typical of those encountered by undergraduate physics majors. Some of the material in Chapters 4–11 will likely be unfamiliar to many readers. However, an attempt has been made to include sufficient details and references, so interested physics and engineering majors should be able to follow the discussions.

At a number of places in the text, I have indicated the source of a mathematical relation in footnotes using the following notation.

AS   M. Abramowitz & I. Stegun (eds.), *Handbook of Mathematical Functions*, Dover Publications, 1972.

CRC   S. Selby (ed.), *Standard Mathematical Tables*, 14th ed., The Chemical Rubber Company, 1965.

GR   I. Gradshteyn & I. Ryzhik, *Table of Integrals, Series and Products*, Academic Press, 1980.

I would like to thank several of my former colleagues, especially Gerry Morgan, Steve Kahn, Bob Palmer, Juan Gallardo, and Scott Berg, for many interesting discussions concerning magnetic fields and the methods used for calculating them. I would like to thank Peter Wanderer and Animesh Jain of the Superconducting Magnet Division at Brookhaven National Laboratory for providing the image of field lines in a RHIC dipole, which has been used on the cover of this book. I would also like to thank Simon Capelin and the staff at Cambridge University Press for their collaboration on this project. Finally, I would like to thank my wife Ruth for her constant support and encouragement.

1

Basic concepts

Magnetic phenomena have been known since antiquity when a natural ore later called lodestone was discovered to attract bits of iron. The scientific study of magnetism dates from around 1600, when William Gilbert summarized experiments on the subject in his treatise *De Magnete*.[1] However, interest in the subject greatly increased after 1820, when Hans Christian Øersted reported that electrical currents could deflect magnetic needles, thereby establishing a connection between the subjects of electricity and magnetism.[2] Almost immediately, André-Marie Ampère, Jean-Baptiste Biot, and Félix Savart performed a series of seminal experiments that determined the forces acting between current loops. Experimental work and theoretical developments continued throughout the first half of the nineteenth century. A long program of experimental investigations by Michael Faraday lead him to the conception that the force between current loops occurred through the action of an intermediary field that existed in the space around the loops. Faraday's field concept was developed mathematically by William Thomson (later Lord Kelvin). This work culminated in a synthesis of knowledge about electrical and magnetic phenomena by James Clerk Maxwell in his famous treatise of 1873. Many clarifications of Maxwell's ideas and studies of their implications were carried out over the next twenty years by a small group of followers. Of particular note was the work of Oliver Heaviside who introduced the use of vector analysis and reworked the set of equations in Maxwell's treatise to the four equations we use today.[3] The resulting Maxwell equations are now accepted as the theoretical description underlying electromagnetic phenomena.

Magnetostatics is the study of the fields, forces, and energy associated with steady currents and magnetic materials. In this chapter, we will review some basic concepts underlying magnetic effects due to conductor currents in free space.

1.1 Current

Experiments have shown that there exist two kinds of electrical charge q, which are denoted as positive and negative. A current I exists when there is a net temporal flow of charge across some arbitrary plane in space.

$$I = \frac{dq}{dt}. \tag{1.1}$$

If the current is flowing through a conductor with length L and cross sectional area A, we can write the current as

$$I = \frac{\rho L A}{L/v} = \rho v \, A,$$

where ρ is the charge density and v is the velocity of the charges. The current density J along some direction n is a vector given by

$$\vec{J} = \frac{I}{A}\hat{n} = \rho \, \vec{v}, \tag{1.2}$$

where $\hat{n}$ is the unit vector perpendicular to A.

If we consider a volume of space V enclosed by a surface S, the conservation of charge requires that any change in the charge density inside V must be compensated by a flow of current through the surface or

$$-\int \frac{\partial \rho}{\partial t} dV = \int \vec{J} \cdot \hat{n} \, dS.$$

Using the Gauss divergence theorem,[1] the right-hand side can be written as

$$\int \vec{J} \cdot \hat{n} \, dS = \int \nabla \cdot \vec{J} \, dV.$$

Then, since V is arbitrary, we can remove the integrands from the volume integrals on both sides of the equation and obtain the continuity equation

$$\frac{\partial \rho}{\partial t} + \nabla \cdot \vec{J} = 0. \tag{1.3}$$

In magnetostatics, we have by definition $\partial \rho / \partial t = 0$, which leads to the relation

$$\nabla \cdot \vec{J} = 0. \tag{1.4}$$

---

[1] Readers unfamiliar with vector analysis should review Appendix B.

Often we are interested in the current flow in a "central" region far from the ends of a magnet. If the current and the geometry are uniform along z in this region, we can simplify the analysis by examining problems in two dimensions. If we consider a conductor whose thickness is small compared with the distance to the observation point, we can approximate the conductor as a *current sheet*.

In addition, we frequently consider line currents or "filaments," where we ignore the transverse dimensions of the conductor altogether and use the equivalent current

$$I = \int \vec{J} \cdot \hat{n} \, dS.$$

1.2 Magnetic forces

Experiments have shown that test currents and charges in the vicinity of a current-carrying conductor experience a force. We assume that this force takes place through the actions of an intermediary magnetic field. The mathematical description of a field is a continuous function that is defined for all points in space and for all times. However, the magnetic field also has physical properties associated with it, such as stored energy. The force experiments can be explained by assuming that a current produces a vector field B, and then this field produces a force on other currents and charges. The vector field B is called the *magnetic flux density*[2] or magnetic field for short. The *magnetic flux* through some surface S is defined as

$$\Phi_B = \int \vec{B} \cdot \vec{dS}. \tag{1.5}$$

The direction of the magnetic field is often represented using Faraday's concept of *lines of induction*.[3][4] The lines of induction are defined to be tangent to the magnetic field at every point in space. It follows that corresponding components of the lines of induction and the magnetic field are always proportional to each other. If ds is a small displacement along the line of induction, we have

$$\frac{dx}{B_x} = \frac{dy}{B_y} = \frac{dz}{B_z} = \frac{ds}{B}.$$

In two Cartesian dimensions, the lines can be plotted, for example, by integrating the equations

---

[2] The vector B is also known as the magnetic induction.
[3] Historically, these curves have been referred to as lines of force.

$$dx = \frac{B_x(x,y)}{B(x,y)}ds$$

$$dy = \frac{B_y(x,y)}{B(x,y)}ds.$$

The magnitude of the magnetic field can be represented by the density of lines in a given region. The lines of induction do not have to form closed loops.[5, 6] In particular, the lines become undefined at locations where $B = 0$.

Now consider two circuits carrying currents I_a and I_b. The force exerted by circuit a on circuit b is found experimentally to be

$$\vec{F}_{ab} = \frac{\mu_0}{4\pi} I_a I_b \oiint \vec{dl_b} \times \frac{\vec{dl_a} \times \vec{R}}{R^3}, \tag{1.6}$$

where the constant $\mu_0 = 4\pi\,10^{-7}$ is known as the *permeability of free space*,[4] dl is a displacement along the circuit in the direction of the current, and R is the distance vector from dl_a to dl_b. Note that the force is proportional to the product of the currents times a geometric factor that depends on the shape and orientations of the two circuits. It is possible to rewrite this equation in a form that manifestly obeys Newton's Third Law of motion. Using the vector triple product identity from Equation B.1 in Appendix B, we have

$$\vec{dl_b} \times (\vec{dl_a} \times \vec{R}) = \vec{dl_a}(\vec{dl_b} \cdot \vec{R}) - \vec{R}(\vec{dl_a} \cdot \vec{dl_b}).$$

The double integral of the first term on the right-hand side is then

$$\oiint \frac{\vec{dl_a}(\vec{dl_b} \cdot \vec{R})}{R^3} = \oint \vec{dl_a} \oint \frac{(\vec{dl_b} \cdot \vec{R})}{R^3} = \oint \vec{dl_a} \oint \frac{dR}{R^2}.$$

The last integral vanishes because the scalar integrand is taken over a closed path. Thus we can express the force as

$$\vec{F}_{ab} = -\frac{\mu_0}{4\pi} I_a I_b \oiint \frac{\vec{R}(\vec{dl_a} \cdot \vec{dl_b})}{R^3}. \tag{1.7}$$

In this form, we see that Newton's law $F_{ab} = -F_{ba}$ is obeyed since R changes direction for the two cases.

Returning to Equation 1.6, we rewrite the force on circuit b in a form that explicitly depends on the current in circuit b and on an integration of the elemental

---

[4] We will use SI units exclusively in this book. For more details, see Appendix A.

interactions taking place around that circuit. We collect the other factors in Equation 1.6 into a new vector B_a, which we define as the magnetic field due to circuit a. Then the force on the circuit can be written as

$$\overrightarrow{F}_{ab} = I_b \oint \overrightarrow{dl}_b \times \overrightarrow{B}_a . \tag{1.8}$$

The force acts at right angles to the direction of B_a. Dropping the subscripts, we see that the force on a charge q moving with velocity v can be written as

$$\overrightarrow{F} = \int \frac{dq}{dt} \overrightarrow{dl} \times \overrightarrow{B} = q \overrightarrow{v} \times \overrightarrow{B} . \tag{1.9}$$

Note that the force only acts on moving charges.

Now consider a rectangular current loop with length L and width w in a constant magnetic field B, as shown in Figure 1.1. The forces on each pair of opposite sides cancel, so there is no net force on the loop. However, there are moment arms between sides 1 and 3 and the axis of the loop. This creates a torque given by

$$\overrightarrow{\tau} = \overrightarrow{r} \times \overrightarrow{F} .$$

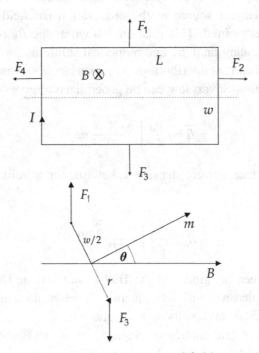

Figure 1.1 Rectangular current loop in an external field.

For the example here,

$$\tau = 2\frac{w}{2}NILB \sin\theta,$$

where N is the number of turns in the loop. We define the *magnetic moment* **m** of a planar loop to lie along the normal n to the loop, so that

$$\vec{m} = NIA\,\hat{n}, \tag{1.10}$$

where A is the area of the loop. Then the torque acting on the loop can be expressed as

$$\vec{\tau} = \vec{m} \times \vec{B}. \tag{1.11}$$

1.3 The Biot-Savart law

Comparing Equations 1.6 and 1.8, we see that the force experiments require that the magnetic field can be expressed in the form

$$\vec{B} = \frac{\mu_0}{4\pi}I \oint \frac{\vec{dl} \times \vec{R}}{R^3}, \tag{1.12}$$

where we have dropped the subscripts referring to circuit a. The vector R points from the current element source to the observation (or field) point where the magnetic field is determined. This relation, known as the *Biot-Savart Law*, is an important tool for finding analytic and numerical solutions for the magnetic field produced by known current distributions. For a surface distribution of current, the total current in the Biot-Savart law can be generalized to give

$$\vec{B} = \frac{\mu_0}{4\pi} \int \frac{\vec{K} \times \vec{R}}{R^3}\,dS, \tag{1.13}$$

where K is the surface current density. Likewise, for a volume distribution of current, we have

$$\vec{B} = \frac{\mu_0}{4\pi} \int \frac{\vec{J} \times \vec{R}}{R^3}\,dV. \tag{1.14}$$

It is important to keep in mind that the Biot-Savart law and many of the other mathematical laws that we will subsequently develop ultimately depend on the validity of the experimental results on magnetic forces.

We consider next several elementary applications of the Biot-Savart law that we will need to refer to later in this book.

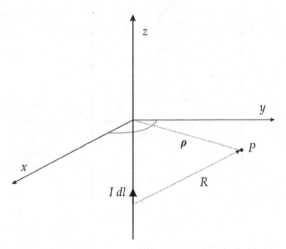

Figure 1.2 Current in a long straight wire.

Example 1.1: field from an infinitely long straight wire

Consider an infinitely long straight wire lying along the z axis, as shown in Figure 1.2. Because of the symmetry, we use cylindrical coordinates. Since the wire is infinitely long, we can chose an observation point P in the plane with $z = 0$ without loss of generality. Since

$$\vec{dl} = dz\,\hat{z}$$
$$\vec{R} = \rho\,\hat{\rho} + z\,\hat{z}$$
$$R = \sqrt{\rho^2 + z^2},$$

the field at point P due to the current in the wire is

$$\vec{B} = \frac{\mu_0}{4\pi}\rho\,\hat{\phi}\,2\,\mathbb{I},$$

where[5]

$$\mathbb{I} = \int_0^\infty \frac{dz}{\{\rho^2 + z^2\}^{3/2}} = \frac{1}{\rho^2}.$$

Thus the magnetic field due to the current in the wire is

$$\vec{B} = \frac{\mu_0 I}{2\pi\rho}\,\hat{\phi}. \tag{1.15}$$

The field is directed azimuthally around the wire and falls off with distance like $1/\rho$.

---

[5] GR 2.271.5.

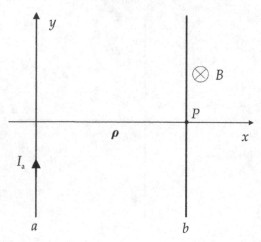

Figure 1.3 Force between two parallel wires.

Example 1.2:   force between two parallel wires

Consider two infinitely long parallel wires, as shown in Figure 1.3. From Equation 1.8, the incremental force between the two wires is

$$\overrightarrow{dF}_b = I_b \,\overrightarrow{dl}_b \times \overrightarrow{B}_a$$

and from the previous example, the field at P due to the current in wire a is

$$\overrightarrow{B}_a = -\frac{\mu_0 I_a}{2\pi \rho} \hat{z}.$$

If the current direction in wire b can be either parallel or antiparallel to the current in wire a, we find that the force per unit length of the wire is

$$\frac{\overrightarrow{dF}_b}{dy} = \pm \frac{\mu_0}{2\pi \rho} I_a I_b \,\hat{x}. \tag{1.16}$$

The force between the wires is attractive when the currents are in the same direction and repulsive when they are antiparallel.

Example 1.3:   field above an infinite current sheet

Consider an infinite current sheet with current flowing uniformly in the y direction. We calculate the magnetic field at point P, shown in Figure 1.4. We have

$$\overrightarrow{K} = K_y \,\hat{y}$$
$$\overrightarrow{R} = x \,\hat{x} + y \,\hat{y} + z_o \,\hat{z}.$$

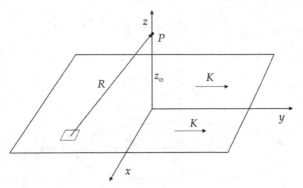

Figure 1.4 Field above an infinite current sheet.

The field is given by

$$\vec{B} = \frac{\mu_0 K_y}{4\pi} \int_{-\infty}^{\infty} \int_{-\infty}^{\infty} \frac{z_o\,\hat{x} - x\,\hat{z}}{\{x^2 + y^2 + z_o^2\}^{3/2}}\,dx\,dy$$

$$= \frac{\mu_0 K_y}{4\pi}(z_o\,\hat{x}\mathbb{I}_1 - \hat{z}\mathbb{I}_2),$$

where

$$\mathbb{I}_1 = \int_{-\infty}^{\infty} \int_{-\infty}^{\infty} \frac{1}{\{x^2 + y^2 + z_o^2\}^{3/2}}\,dx\,dy = \frac{2\pi}{z_o}$$

and

$$\mathbb{I}_2 = \int_{-\infty}^{\infty} \int_{-\infty}^{\infty} \frac{x}{\{x^2 + y^2 + z_o^2\}^{3/2}}\,dx\,dy = 0.$$

The integral $\mathbb{I}_2$ vanishes because the integrand is an odd function and the integration extends over an even interval. The magnetic field above the sheet is

$$\vec{B} = \frac{\mu_0}{2} K_y\,\hat{x}.$$

The direction of the field is parallel to the sheet and perpendicular to the current density. The magnitude of the field is constant and independent of the distance from the sheet. In the general case, the field above the sheet can be written as

$$\vec{B} = \frac{\mu_0}{2}\vec{K} \times \hat{n}, \tag{1.17}$$

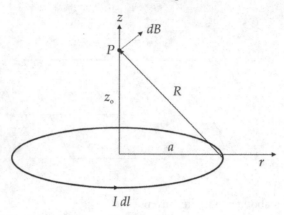

Figure 1.5 Field along the axis of a current loop.

where n is the normal to the sheet pointing to the side where B is computed. Note that the direction of B follows the right-hand rule with respect to the current filaments in the sheet.

Example 1.4:   on-axis field due to a circular current loop
We look for the field at a point P that is along the axis of the loop and a distance z_o above the plane of the current loop, as shown in Figure 1.5. In cylindrical coordinates, we have

$$\vec{dl} = a\,d\phi\,\hat{\phi}$$
$$\vec{R} = -a\,\hat{r} + z_o\,\hat{z}.$$

The contributions of the current elements to the field at P lie in a cone surrounding P. By symmetry, the net field must be in the z direction and

$$(\vec{dl} \times \vec{R})_z = a^2\,d\phi.$$

Thus we have

$$\begin{aligned}
B_z &= \frac{\mu_0 I}{4\pi} \int_0^{2\pi} \frac{a^2}{\{a^2 + z_o^2\}^{3/2}}\,d\phi \\
&= \frac{\mu_0 I\,a^2}{2\{a^2 + z_o^2\}^{3/2}}.
\end{aligned}$$

(1.18)

Note that B_z is proportional to the area of the current loop and falls off at large distances like z_o^{-3}. The field is largest at the center of the loop where the value is

$$B_{z0} = \frac{\mu_0 I}{2a}.$$

1.4 Divergence of the magnetic field

Starting from the Biot-Savart law,

$$\vec{B} = \frac{\mu_0}{4\pi} \int \frac{\vec{J} \times \vec{R}}{R^3} \, dV$$

and taking the divergence of both sides, we have

$$\nabla \cdot \vec{B} = \frac{\mu_0}{4\pi} \int \frac{\nabla \cdot (\vec{J} \times \vec{R})}{R^3} \, dV',$$

where we use primes to indicate the use of source coordinates. The operator ∇ is defined in terms of field coordinates, while the distance R is a function of both source and field coordinates. Using the vector identity B.4, we can write

$$\nabla \cdot (\vec{J} \times \vec{R}) = \vec{R} \cdot (\nabla \times \vec{J}) - \vec{J} \cdot (\nabla \times \vec{R}).$$

The first term on the right-hand side vanishes because $\nabla \times \vec{J} = 0$. When the second term is written in terms of a determinant, we obtain

$$\nabla \times \vec{R} = \begin{vmatrix} \hat{x} & \hat{y} & \hat{z} \\ \partial_x & \partial_y & \partial_z \\ x - x' & y - y' & z - z' \end{vmatrix} = 0.$$

Thus we find that the divergence of the magnetic field vanishes.

$$\nabla \cdot \vec{B} = 0. \tag{1.19}$$

This vector relation is one of the fundamental properties of magnetic fields. Its validity depends on the fact that isolated magnetic charges (monopoles) do not appear to exist.

If we integrate Equation 1.19 over some volume of space V that is enclosed by a surface S, we get

$$\int \nabla \cdot \vec{B} \, dV = 0.$$

Then using the divergence theorem, we find *Gauss's Law for magnetism*

$$\int \vec{B} \cdot \hat{n} \, dS = 0. \tag{1.20}$$

1.5 Circulation of the magnetic field

Returning again to the Biot-Savart law, consider the integral

$$\mathbb{I} = \int \frac{\overrightarrow{J'} \times \overrightarrow{R}}{R^3} dV'$$
$$= \int \left(\nabla \left(\frac{1}{R} \right) \times \overrightarrow{J'} \right) dV'.$$

Using the vector identity B.6, we can write this as

$$\mathbb{I} = \int \nabla \times \frac{\overrightarrow{J'}}{R} dV' - \int \frac{1}{R} \nabla \times \overrightarrow{J'} dV'.$$

The quantity $\nabla \times \overrightarrow{J'} = 0$ in the second integral. We can bring the ∇ operator in the first term outside the integral sign because it operates on the observation point coordinates, while the integral is over the source point coordinates. Thus

$$\mathbb{I} = \nabla \times \int \frac{\overrightarrow{J'}}{R} dV'$$

and an alternate expression for the magnetic field is

$$\overrightarrow{B} = \frac{\mu_0}{4\pi} \int \nabla \times \frac{\overrightarrow{J'}}{R} dV'. \tag{1.21}$$

Taking the curl of both sides of this equation, we find

$$\nabla \times \overrightarrow{B} = \frac{\mu_0}{4\pi} \nabla \times \nabla \times \int \frac{\overrightarrow{J'}}{R} dV'.$$

Using the vector relation B.7, we can write this in the form

$$\nabla \times \overrightarrow{B} = \frac{\mu_0}{4\pi} \left[\nabla \int \nabla \cdot \left(\frac{\overrightarrow{J'}}{R} \right) dV' - \int \nabla^2 \left(\frac{\overrightarrow{J'}}{R} \right) dV' \right].$$

Then since ∇ does not operate on J', we have

$$\nabla \times \overrightarrow{B} = \frac{\mu_0}{4\pi} \left[\nabla \int \overrightarrow{J'} \cdot \nabla \left(\frac{1}{R} \right) dV' - \int \overrightarrow{J'} \nabla^2 \left(\frac{1}{R} \right) dV' \right]. \tag{1.22}$$

Consider for the moment the relation

$$\nabla \left(\frac{1}{R} \right) = -\nabla' \left(\frac{1}{R} \right).$$

In the second integral in Equation 1.22 when $R \neq 0$, we have

$$\nabla^2 \left(\frac{1}{R}\right) = \frac{1}{r^2} \partial_r \left[r^2 \partial_r \left(\frac{1}{R}\right)\right]$$

in spherical coordinates. Then since $R = |\vec{r} - \vec{r}'|$, we find that $\nabla^2 \left(\frac{1}{R}\right) = 0$. Although $\nabla^2 \left(\frac{1}{R}\right)$ is undetermined when $R = 0$, the integral of this expression is still defined. Performing the integral on a small sphere surrounding $R = 0$, we find

$$\int \nabla^2 \left(\frac{1}{R}\right) dV = \int \nabla \cdot \nabla \left(\frac{1}{R}\right) dV = \int \nabla \left(\frac{1}{R}\right) \cdot \vec{dS}$$

using the divergence theorem. Evaluating the last integral on the surface of the small sphere, we find that

$$\int \nabla \left(\frac{1}{R}\right) \cdot \vec{dS} = -\frac{1}{R^2} 4\pi R^2 = -4\pi.$$

We can summarize these results by writing the expression in terms of the Dirac delta function δ.

$$\nabla^2 \left(\frac{1}{R}\right) = -4\pi\delta(R). \tag{1.23}$$

Now we can do the first integral in Equation 1.22 using Equation B.3 to give

$$\int \vec{J}' \cdot \nabla' \left(\frac{1}{R}\right) dV' = \int \nabla' \cdot \frac{\vec{J}'}{R} dV' - \int \frac{1}{R} \nabla \cdot \vec{J}' dV'.$$

The first term on the right-hand side can be converted to a surface integral using the divergence theorem. It vanishes if the surface enclosing the volume in the integrals is sufficiently large. The second integral also vanishes because $\nabla' \cdot \vec{J}' = 0$ for magnetostatics. Thus we are only left with the second integral in Equation 1.22, which because of the delta function from Equation 1.23, gives

$$\nabla \times \vec{B} = \mu_0 \vec{J}. \tag{1.24}$$

Thus we have shown that a steady current creates a magnetic field that circulates around the current. This is a second fundamental vector relation for magnetic fields.[6]

---

[6] We will find in Chapter 10 that this relation requires an additional term if the current varies with time.

1.6 The Ampère law

If we integrate both sides of Equation 1.24 over an arbitrary surface S, we find

$$\int (\nabla \times \vec{B}) \cdot \vec{dS} = \mu_0 \int \vec{J} \cdot \vec{dS}.$$

On the right side, the current density integrated over the surface gives the total current I. On the left side, we can use Stokes's theorem from Appendix B to give

$$\int (\nabla \times \vec{B}) \cdot \vec{dS} = \oint \vec{B} \cdot \vec{dl},$$

where the contour on the right-hand side extends along the perimeter of the surface S. Thus we have the result

$$\oint \vec{B} \cdot \vec{dl} = \mu_0 I. \tag{1.25}$$

This equation is known as the Ampère law.[7] It can be most usefully applied in highly symmetric cases where, for example, the magnitude of the field is constant along the integration path.

Again let us consider several elementary examples of using the Ampère law to derive results that we use later in the book.

Example 1.5: a long cylindrical conductor
Consider a long cylindrical conductor with constant current density J inside the radius a, as shown in Figure 1.6. Since by symmetry the magnitude of the field must be independent of ϕ, we choose a circular path of integration. When the path is outside the conductor, all of the current is enclosed by the path and the Ampère law gives $B_\phi 2\pi\rho = \mu_0 I$. Thus the field outside the conductor is

$$B_\phi = \frac{\mu_0 I}{2\pi\rho}, \tag{1.26}$$

which falls off like $1/\rho$. Since this result is independent of the radius of the conductor, it also applies to the field from a current filament, which we previously derived using the Biot-Savart equation.

When the path of integration is inside the conductor, only part of current is enclosed by the path and the Ampère law gives

$$B_\phi 2\pi\rho = \mu_0 \frac{\pi\rho^2}{\pi a^2} I.$$

---

[7] According to O. Darrigol,[2] this equation was first given by Maxwell. In that case, we agree with him that it's really not appropriate to call it Ampère's law.

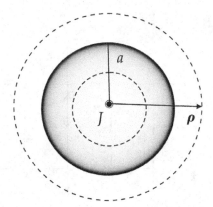

Figure 1.6 Cylindrical conductor of radius a. The two integration paths are shown with dotted lines.

Thus the field inside the conductor is

$$B_\phi = \frac{\mu_0 I \rho}{2\pi a^2} = \frac{\mu_0 J}{2}\rho, \tag{1.27}$$

which increases linearly with ρ.

Example 1.6: ideal solenoid
We define an ideal solenoid as an infinitely long system of parallel circular current loops with radius a, as shown in Figure 1.7. This is an approximation to a real solenoid when the observation points are far from the ends of the solenoid and the conductor is tightly wound, such that we can ignore any gaps or the helical nature of the windings. First consider a cylinder containing the points *achj*. From Gauss's law, Equation 1.20, we know that the flux passing through the surface must be 0. From symmetry, the flux through the top and through the bottom faces of the cylinder have to cancel. The contribution through the side of the cylinder then gives

$$2\pi \rho L B_\rho = 0.$$

Since the same argument applies for a cylinder of any radius, we must have $B_\rho = 0$ everywhere for the ideal solenoid.

Now consider the Ampère law applied to the path *bdgi*. The contributions to the integral vanish along *bd* and *gi* since $B_\rho = 0$. Then we have

$$B_z(0)L - B_z(r_1)L = \mu_0 n I L,$$

where n is the number of conductor turns per unit length, or that

$$B_z(r_1) = B_z(0) - \mu_0 n I.$$

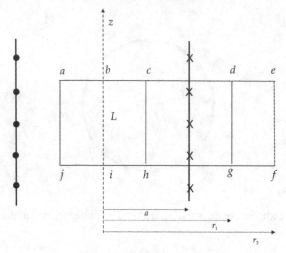

Figure 1.7 Cross-section of an ideal solenoid. The dashed line is the axis of the solenoid. The dots and crosses refer to the direction of the current.

If we apply the same argument over the path *befi*, we find that

$$B_z\left(r_2\right) = B_z\left(0\right) - \mu_0\, n\, I.$$

Both of these equations for B_z outside the solenoid have the same right-hand side. Thus the value of B_z outside the solenoid is constant, independent of radius. However, if we consider the total flux outside the solenoid, we find

$$\Phi_B = \int_0^{2\pi}\int_a^\infty B_z^{out}\, r\, dr\, d\phi.$$

The total flux would be infinite if B_z outside the solenoid is any constant other than 0. This is clearly non-physical, so we must have $B_z = 0$ everywhere outside the solenoid.

Since the field vanishes outside the solenoid, applying the Ampère law to the path *bdgi* gives

$$B_z\left(0\right) = \mu_0 n I.$$

Similarly, on the path *cdgh* we find

$$B_z\left(r\right) = \mu_0 n I, \tag{1.28}$$

where we write r for the length *bc*. Thus the field of the ideal solenoid is constant and along the axis of the solenoid on the inside and it vanishes outside.

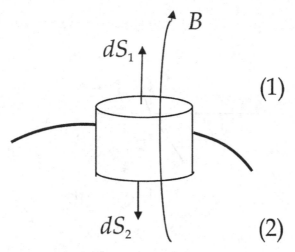

Figure 1.8 Gaussian pillbox across a current sheet.

1.7 Boundary conditions at a current sheet

We now consider how the magnetic field is influenced by the presence of a current sheet. Assume we have a current sheet, as shown in Figure 1.8. We construct a cylindrical pillbox across the sheet with an infinitesimal height along the normal to the surface. Then applying Equation 1.20, we find that

$$B_{1n} S - B_{2n} S = 0.$$

Since the surface S is arbitrary, it follows that $B_{1n} = B_{2n}$ or that

$$(\vec{B_2} - \vec{B_1}) \cdot \hat{n} = 0. \tag{1.29}$$

Thus the normal component of B must be continuous across a current sheet.

Next construct a closed path across the current sheet, as shown in Figure 1.9. Assume the path length perpendicular to the surface is infinitesimally small. The path encloses any current present in the sheet. Applying the Ampère law, we find

$$-B_{1t} L + B_{2t} L = \mu_0 K L.$$

Thus the change in the field across the sheet is

$$B_{2t} - B_{1t} = \mu_0 K \tag{1.30}$$

or in general

$$(\vec{B_2} - \vec{B_1}) \times \hat{n} = \mu_0 \vec{K}. \tag{1.31}$$

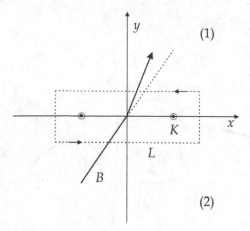

Figure 1.9 Fields near a current sheet.

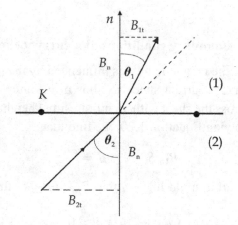

Figure 1.10 Refraction of the magnetic field crossing a current sheet.

The tangential component of B changes by an amount proportional to the current density when crossing the sheet.

Finally, let us consider the angles between the magnetic field vectors and the normal to the current sheet, as shown in Figure 1.10. In region (2), the magnetic field vector makes an angle with the normal to the current sheet given by

$$\tan \theta_2 = \frac{B_{2t}}{B_n}.$$

The corresponding angle with the normal in region (1) is given by

$$\tan \theta_1 = \frac{B_{1t}}{B_n}$$

$$= \frac{B_{2t} - \mu_0 K}{B_n}$$

$$= \tan \theta_2 - \frac{\mu_0 K}{B_n}.$$

We see that in crossing the current sheet, the vector B is refracted in the direction of the field from the current sheet, i.e., toward the normal for the positive current density K shown here.

1.8 Inductance

Consider a coil with N turns. We define the *flux linkage* to be the product of the magnetic flux going through the coil multiplied by the number of turns. The flux linkage is proportional to the current flowing through the coils. We define the coefficient of proportionality to be the *self-inductance L* of the coil. Thus we have

$$L = \frac{N \Phi_B}{I}. \tag{1.32}$$

Example 1.7:   self-inductance of an ideal solenoid

For an ideal solenoid with N turns in a length d and radius R, the field from Equation 1.28 is

$$B_z \simeq \frac{\mu_0 N I}{d}.$$

The flux in the solenoid is

$$\Phi_B = \frac{\mu_0 N I}{d} \pi R^2,$$

so the self-inductance is

$$L = \frac{\mu_0 N^2}{d} \pi R^2. \tag{1.33}$$

This result ignores any end effects present in a real solenoid.

If we now consider two coils, the *mutual inductance M* is defined as the flux linkage in the second coil due to the current in the first coil. Thus we have

$$M = \frac{N_2 \Phi_{2,1}}{I_1}, \qquad (1.34)$$

where $\Phi_{2,1}$ is the flux in coil 2 due to the current in coil 1. In general, M can be defined using the Neumann equation [7]

$$M = \frac{\mu_0}{4\pi} \iint \frac{\overrightarrow{dl_1} \cdot \overrightarrow{dl_2}}{r_{12}}, \qquad (1.35)$$

where r_{12} is the distance from the current element in the first coil to the current element in the second coil. Note that this shows that M is a constant times a geometric factor. The symmetry of this equation between the two coils shows that M of coil 2 due to current in coil 1 is the same as M for coil 1 due to current in coil 2.

Example 1.8:   mutual inductance of two coaxial solenoids
Assume we have two coaxial solenoids. The first solenoid has length d_1 and both solenoids have approximately the same radius R. Then

$$\Phi_{2,1} = \frac{\mu_0 \, N_1 \, I_1}{d_1} \pi R^2$$

and the mutual inductance is

$$M = \frac{\mu_0 N_1 N_2}{d_1} \pi R^2. \qquad (1.36)$$

The force between two coaxial coils can be expressed in terms of the derivative of their mutual inductance.[8]

$$F_z = I_1 \, I_2 \frac{\partial M}{\partial z}. \qquad (1.37)$$

1.9 Energy stored in the magnetic field

We can obtain a rough estimate for the energy stored in a magnetic field[8] by considering a simple LR circuit, as shown in Figure 1.11. From Kirchhoff's circuit laws,[9] we have

$$IV = I^2 R + LI \frac{dI}{dt}.$$

---

[8] We will reexamine this question more rigorously in Chapter 10.

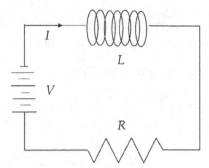

Figure 1.11 LR circuit.

The power $P = IV$ generated by the battery is distributed between the power lost to resistive heating in R and the power in the magnetic field associated with the inductor L. The energy in the magnetic field is then

$$W_B = \int P\, dt = \int LI\frac{dI}{dt} dt = \int LI\, dI$$
$$= \frac{1}{2}LI^2.$$

(1.38)

If we consider the inductor to be a long solenoid, then using Equations 1.28 and 1.33,

$$W_B = \frac{1}{2}\frac{\mu_0 N^2}{d}\,\pi R^2\,\frac{B^2 d^2}{\mu_0^2 N^2} = \frac{1}{2}\frac{B^2}{\mu_0}\,\pi R^2 d$$

and the energy density in the magnetic field is

$$w_B = \frac{B^2}{2\,\mu_0}.$$

(1.39)

References

[1] F. Bonaudi, Magnets in particle physics, in S. Turner (ed.), *CERN Accelerator School: Magnetic Measurements and Alignment*, Montreux, 1992, p. 2–5.
[2] O. Darrigol, *Electrodynamics from Ampère to Einstein*, Oxford University Press, 2000, p. 142.
[3] B. Hunt, *The Maxwellians*, Cornell University Press, 1991, p. 70–71, 245–247.
[4] D. Halliday & R. Resnick, *Physics for Students of Science and Engineering*, 2nd ed., Wiley, 1962, p. 759–760.
[5] K.L. McDonald, Topology of steady current magnetic fields, *Am. J. Phys.* 22:586, 1954.
[6] M. Lieberherr, The magnetic field lines of a helical coil are not simple loops, *Am. J. Phys.* 78:1117, 2010.

[7] G. Harnwell, *Principles of Electricity and Magnetism*, 2nd ed., McGraw-Hill, 1949, p. 322.

[8] P. Lorrain & D. Corson, *Electromagnetic Fields and Waves*, 2nd ed., Freeman, 1970, p. 360.

[9] D. Tomboulian, *Electric and Magnetic Fields*, Harcourt, Brace & World, 1965, p. 267–269.

2

Magnetic materials

We saw in the previous chapter that magnetic fields are produced in a vacuum by currents in conductors. In this chapter, we will consider magnetic effects associated with matter. All materials have spin and orbital motion of charges at the atomic scale. For most materials, the random orientations of these internal currents tend to cancel out significant magnetic effects. However, in certain *magnetic materials*, such as iron or permanent magnets, these internal currents do not cancel, and there is a net external effect that also produces or enhances magnetic fields. In the field theory we have been discussing, the magnetic field B must be a continuous function of position. Thus, in magnetic materials, the macroscopic field B must be an average of the rapidly varying local fields surrounding the atoms in the material.[1]

2.1 Magnetization

When a magnetic material is placed in an external magnetic field, magnetic dipoles in the material set up internal fields that modify the applied field. We saw in Equation 1.10 that a current loop has an associated magnetic moment m. We define the *magnetization* vector M as the average magnetic moment per unit volume

$$\vec{M} = \sum_i \frac{\vec{m}_i}{V} = \frac{N_i I_i A_i}{V} \, \hat{n}_i \,, \tag{2.1}$$

where A_i is the area of loop i. The volume V must be large enough so the sum is statistically significant, yet small enough so that we can treat the variation of M as approximately continuous. For uniform magnetization, all the internal current loops cancel. However, as shown in Figure 2.1, there is still a net current around the surface of the material.

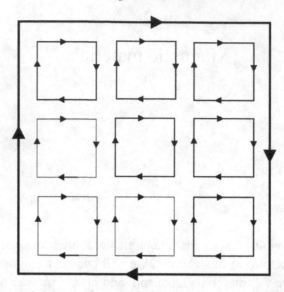

Figure 2.1 Magnetic moment loops in a uniformly magnetized material.

It follows that

$$M = \frac{I_m \sum_i A_i}{AL} = \frac{I_m}{L} = K_m,$$

where I_m is the Amperian loop current and K_m is the magnetization surface current density. In vector terms,

$$\vec{K}_m = \vec{M} \times \hat{n} . \tag{2.2}$$

Now consider the case when there is a nonuniform distribution of magnetization inside the volume, as shown in Figure 2.2. For the current loop on the left we have

$$M_z' = \frac{I'}{\Delta z},$$

while the loop on the right gives

$$M_z'' = M_z' + \frac{\partial M_z}{\partial y}\,\Delta y = \frac{I''}{\Delta z}.$$

Along the line AB, there is a net current

$$\frac{I_m}{\Delta z} = \frac{I'' - I'}{\Delta z} = \frac{\partial M_z}{\partial y}\,\Delta y. \tag{2.3}$$

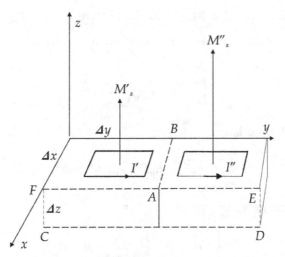

Figure 2.2 Magnetic moment loops in a nonuniform magnetized material.

If we integrate around the front face (*CDEF*), we find

$$\oint \vec{M} \cdot \vec{dl} = \left(M'_z + \frac{\partial M_z}{\partial y} \Delta y \right) \Delta z - M'_z \Delta z = \frac{\partial M_z}{\partial y} \Delta y \, \Delta z.$$

Therefore, using Equation 2.3 we find, in analogy with the Ampère law, that

$$\oint \vec{M} \cdot \vec{dl} = I_m, \qquad (2.4)$$

where I_m is the effective number of internal amp-turns through the loop of integration. Applying Stokes's theorem, we can rewrite Equation 2.4 as

$$\int (\nabla \times \vec{M}) \cdot \vec{dS} = \int \vec{J}_m \cdot \vec{dS}.$$

The Amperian volume current density is then

$$\vec{J}_m = \nabla \times \vec{M}. \qquad (2.5)$$

This implies that the volume current density vanishes for homogeneous materials. When magnetic materials are present, the Biot-Savart law must be generalized to

$$\vec{B} = \frac{\mu_0}{4\pi} \int \frac{(\vec{K} + \vec{K}_m) \times \vec{R}}{R^3} \, dS + \frac{\mu_0}{4\pi} \int \frac{(\vec{J} + \vec{J}_m) \times \vec{R}}{R^3} \, dV,$$

where J and K without a subscript refer to the conduction current densities of free charges.

2.2 Magnetic field intensity

Returning to the Ampère law, we take into account the effect of Amperian currents by writing

$$\oint \vec{B} \cdot \vec{dl} = \mu_0(I + I_m).$$

Using Equation 2.4, we have

$$\oint \vec{B} \cdot \vec{dl} = \mu_0 I + \mu_0 \oint \vec{M} \cdot \vec{dl}.$$

Combining the line integrals gives

$$\oint (\vec{B} - \mu_0 \vec{M}) \cdot \vec{dl} = \mu_0 I.$$

We can define an auxiliary vector H, known as the *magnetic intensity*, as

$$\vec{H} = \frac{\vec{B}}{\mu_0} - \vec{M}, \tag{2.6}$$

so that the magnetic flux density is

$$\vec{B} = \mu_0(\vec{H} + \vec{M}). \tag{2.7}$$

In free space, this reduces to

$$\vec{B} = \mu_0 \vec{H}. \tag{2.8}$$

The vectors B and H describe different aspects of the same magnetic field. The advantage of working with this new vector is that the Ampère law for H

$$\oint \vec{H} \cdot \vec{dl} = I \tag{2.9}$$

only depends on the true conduction current that crosses the path of integration, despite the presence of any magnetic materials. Applying Stokes's theorem to Equation 2.9, we get

$$\int (\nabla \times \vec{H}) \cdot \vec{dS} = \int \vec{J} \cdot \vec{dS}.$$

Since the surface of integration is arbitrary, we find that H satisfies the differential equation

$$\nabla \times \vec{H} = \vec{J}, \tag{2.10}$$

where J is the conduction current density.

We have introduced the vector H here as a useful mathematical artifact to take into account the averaged behavior of atomic currents in matter. However, there is a long history of trying to understand the physical meaning of H and of examining the distinction between vectors B and H.[2, 3]

2.3 Permeability and susceptibility

In homogeneous and isotropic materials, the magnetization is usually found to be proportional to the magnetic intensity, or

$$\vec{M} = \chi \vec{H}, \tag{2.11}$$

where the dimensionless coefficient χ is known as the *susceptibility*. Using this, we find that Equation 2.7 can be rewritten

$$\vec{B} = \mu_0 \vec{H} + \mu_0 \chi \vec{H} = \mu_0 (1 + \chi) \vec{H}.$$

It is useful to define the *permeability* for magnetic materials as

$$\mu = \mu_r \mu_0, \tag{2.12}$$

where μ_r is a dimensionless quantity known as the *relative permeability*. The susceptibility and the relative permeability are related by

$$\mu_r = 1 + \chi, \tag{2.13}$$

and the general relation between B and H can be written as

$$\vec{B} = \mu \vec{H}. \tag{2.14}$$

Materials where the directions of B and H are parallel are called *linear* materials.

2.4 Types of magnetism

Normally the random orientations of atomic orbits and particle spins cause the associated magnetic moments in a material to cancel, so there is no net magnetization. However, when an external magnetic field is applied to the material, the electron orbital velocity increases for one direction of circulation and decreases for the opposite direction. This results in a small net magnetic moment that is present in all materials and is known as *diamagnetism*. The difference in frequency between the two orbital directions can be shown to be

$$\Delta\omega = \pm\frac{eB}{2m_e},$$

where e is the electron charge and m_e is its mass.[4] The resulting net magnetic moment Δm is

$$\Delta m \simeq -\frac{e^2 r^2 B}{4m_e}, \tag{2.15}$$

where r is the radius of the atomic orbit. Note that the induced moment is *opposite* to the direction of the applied magnetic field. This effect is very small and is masked by larger effects in paramagnetic and ferromagnetic materials. Diamagnetic materials have $\chi < 0$ and $\mu_r < 1$, independent of temperature.

In *paramagnetic* materials, the magnetic moments from orbital motion and spins do not cancel, resulting in a small permanent moment. In an external magnetic field, torques tend to align the magnetic moments with the direction of the field. In these materials, the induced fields act to increase the magnitude of the applied field and $\mu_r > 1$. The degree of alignment is decreased by internal collisions, vibrations and thermal agitation inside the material. The resulting magnetization is a spin effect given by the Langevin equation

$$M(H) = Nm\left[\coth\frac{mH}{kT} - \frac{kT}{mH}\right], \tag{2.16}$$

where N is the number of atoms per unit volume, k is Boltzmann's constant, and T is the temperature.[5] Note that the dependence of M on the magnetic intensity H is temperature dependent and nonlinear in general. However, in the case where $\frac{mH}{kT} \ll 1$, the paramagnetic susceptibility is given by Curie's law

$$\chi = \frac{M}{H} = \frac{N m^2}{3kT}. \tag{2.17}$$

In certain crystalline materials where one of the electron shells is not filled, it is possible for one or more electrons to have unbalanced electron spins. In these *ferromagnetic* materials, it is possible to achieve a very high degree of magnetic alignment. Below a characteristic temperature known as the Curie temperature, coupling is possible between neighboring atoms, which can act together in regions known as *domains*. In an external magnetic field, the size of favorably oriented domains can grow. In addition, the magnetization directions in each domain tend to align with the external field. The dependence of B or M on H is very nonlinear in ferromagnetic materials. For example, we show a BH curve for a low-carbon steel alloy in Figure 2.3.[6] The flux density increases extremely rapidly for small values

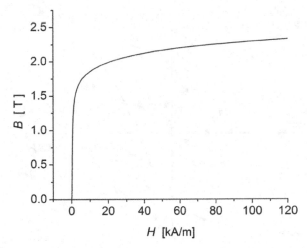

Figure 2.3 **A** *BH* curve for SAE 1020 low-carbon steel.

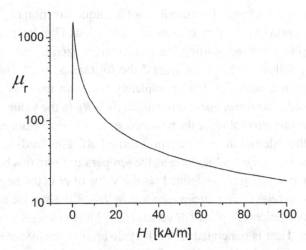

Figure 2.4 A μ*H* curve for SAE 1020 low-carbon steel.

of *H*. Then, as *H* increases further, the magnetization domains begin to saturate, and the curve starts to level out. Finally, for very large applied fields, the magnetization domains become completely saturated, and the growth in *B* is only due to the increase in the conduction current.

The relative permeability in a ferromagnetic material can be much larger than 1, as shown in Figure 2.4. The permeability in this example quickly reaches a maximum value ~1,525 for an excitation of 365 A/m. For larger excitations, the relative permeability decreases steadily.

If the current and thus *H* is cycled up and down in a ferromagnetic material, we find that a plot of *B* versus *H* has the characteristic shape illustrated in Figure 2.5.

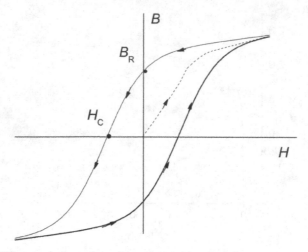

Figure 2.5 Hysteresis loop for a ferromagnetic material.

The material exhibits *hysteresis* because B is not a unique function of H. The value of B depends on the previous history of how H was varied. The dotted line shows the increase in B as H is increased, starting from an unmagnetized sample. After the initial excitation, $B(H)$ follows the arrows around the hysteresis loop. This effect arises because the domain boundaries don't completely return to their previous locations when H is reversed. The *remanence* or *remanent field* B_R is the value of B when H is returned to 0. The nonzero value for the remanence shows that the material can remain magnetic when the external driving current is turned off. This leads to the possibility of making permanent magnets,[1] so long as the temperature remains below the Curie temperature. The *coercivity* H_C is defined[2] as the value of H in the negative direction that is required to get $B = 0$. The *intrinsic coercivity* H_{Ci} is the reverse field required to remove the magnetization in a plot of M versus H.[7] The hysteresis loop is symmetric around the origin. Heat is generated for each cycle around the hysteresis loop.[3]

Magnetic materials with low values of coercivity are designated as "soft." In this case, it is easy for the magnetization to change direction as the external current changes, so these materials are suitable for *ac* operation.[8] A number of soft magnetic materials that have high permeabilities at low values of B are listed in Table 2.1. Also listed are the values of B corresponding to the maximum permeability, the coercivity, and the saturation value of B. The electrical resistivity ρ_e of the material is important for considerations of eddy current[4] losses in time-varying operations. Values for pure iron are also listed for comparison.

---

[1] Permanent magnets are discussed in more detail in Chapter 9.
[2] Historically, this quantity is known as the coercive force.
[3] The energy loss in the hysteresis loop is discussed in Chapter 10.
[4] Eddy currents are discussed in more detail in Chapter 10.

Table 2.1 *Selected soft magnetic alloys [9]*

Alloy	Initial μ_r	Max μ_r	B at max μ_r [T]	H_c [Oe]*	B_{sat} [T]	ρ_e [$\mu\Omega$-cm]
Sinimax	2,200	50,000	0.54	0.06	1.10	90
Monimax	3,000	60,000	0.62	0.06	1.45	80
16 Alfenol	4,000	80,000	0.35	0.044	0.80	153
Mumetal	20,000	100,000	0.20	0.30	0.65	60
1040 alloy	20,000	100,000	0.20	0.20	0.60	56
Supermalloy	55,000	300,000	0.40	0.006	0.68–0.78	65
Iron	150	5,000	0.80	1.00	2.14	10

* 1 Oe = 1 10^{-4} T/μ_0

2.5 Magnetic circuits

Ferromagnetic materials tend to concentrate magnetic flux. The permeability of iron can be thousands of times larger than that of free space. When analyzing "circuits" made up of coils and pieces of ferromagnetic materials, it is common to assume that all the flux goes through and is uniformly distributed inside the ferromagnetic material. A coil sets up Amperian currents in the material near the coil that continue to initiate further currents along the material.[10] Assume that we have an iron ring that is energized with a coil, as shown in Figure 2.6. From the Ampère law for H (Equation 2.9), we find that B inside the ring is

$$B_\phi = \frac{\mu N I}{2\pi R},$$

and the magnetic flux inside the ring is

$$\Phi_B = \left(\frac{\mu \pi r^2}{2\pi R}\right) N I. \tag{2.18}$$

This expression is only approximate because it ignores any leakage of the flux from the ring. Equation 2.18 resembles Ohm's law for circuits

$$I = \frac{V}{R_e},$$

where NI corresponds to the driving voltage and the resulting flux corresponds to the current. The term "analogous to the electrical resistance" R_e is called the *reluctance*, which we see can be expressed as

$$\mathcal{R} - \frac{L}{\mu A}, \tag{2.19}$$

where L is the path length in the material, and A is its cross-sectional area.

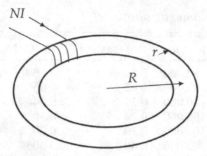

Figure 2.6 Magnetic circuit (Rowland ring).

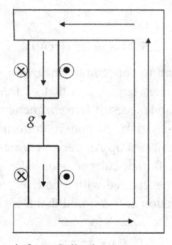

Figure 2.7 Magnetic circuit for a *C*-dipole magnet.

Example 2.1:   C-shaped electromagnet
Consider an electromagnet with an air gap, as shown in Figure 2.7. From the
Ampère law

$$N I = H_i L_i + H_g L_g, \qquad (2.20)$$

where L is the mean length through the region, and the subscripts i and g refer to the
iron and the gap. On the assumption there is no leakage flux, we have $B_i A_i = B_g A_g$.
Substituting into Equation 2.20, we get

$$N I = \frac{B_i}{\mu} L_i + \frac{B_g}{\mu_0} L_g = \frac{B_g A_g}{A_i} \frac{L_i}{\mu} + \frac{B_g}{\mu_0} L_g,$$

so that

$$N I = B_g A_g \left[\frac{L_i}{\mu A_i} + \frac{L_g}{\mu_0 A_g} \right].$$

Since $\mu \gg \mu_0$, we can neglect the reluctance in the iron for reasonable values of L and A and find that the field in the gap is

$$B_g \simeq \frac{\mu_0 N I}{L_g}.$$ (2.21)

2.6 Boundary conditions between regions with different μ

We now consider the constraints on the magnetic field at the boundary between two regions having different permeabilities. The pillbox construction in Figure 1.8 still applies in this case, so

$$B_{1n} = B_{2n}.$$ (2.22)

However, when dealing with permeable materials, Equation 1.30 for the tangential component of B is no longer accurate. In the present case, we can use the Ampère law for H for a path that encompasses both sides of the boundary to find that

$$H_{2t} - H_{1t} = K,$$ (2.23)

where we recall that K is the surface current density. If there is no surface current at the boundary, then the tangential component of H is continuous.

Consider a linear material with $K = 0$ and let θ be the angle between the magnetic field and the normal to the surface, as shown in Figure 2.8. Then we have from Equations 2.22 and 2.23

$$\mu_1 H_1 \cos \theta_1 = \mu_2 H_2 \cos \theta_2$$
$$H_1 \sin \theta_1 = H_2 \sin \theta_2.$$

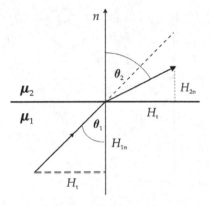

Figure 2.8 Refraction of H at the boundary between two permeable materials.

Dividing these two equations, we find that

$$\frac{\tan \theta_1}{\tan \theta_2} = \frac{\mu_1}{\mu_2}. \tag{2.24}$$

If the field is incident from region 1 and μ is much larger in region 2 than region 1, then θ_1 will be smaller than θ_2. Thus, the field will make a larger angle with respect to the normal in the region with larger μ.

As a special case, consider the boundary between vacuum in region 1 and iron in region 2 when $K = 0$. Then, using Equation 2.23, we find

$$\frac{B_{t2}}{B_{t1}} = \frac{\mu_2}{\mu_1}.$$

It is often useful to make the approximation that μ_2 in the iron is infinite. Then the right-hand side of this equation is infinite, which demands that $B_{t1} = 0$. In this case, the field in the vacuum region would be perpendicular to the iron surface.

2.7 Method of images

Some magnetostatic problems with planar or spherical boundaries can be solved using the method of images.[11] In this method, the presence of an iron boundary is replaced with virtual currents, which, together with the currents from true conductors, reproduce the correct boundary conditions. Consider a current I in a region 1 with permeability μ_1 a distance d from the planar boundary with a region 2 of material with permeability μ_2, as shown in Figure 2.9. Let us designate case (a) to be the situation when the observation point P is in region 1. Then, to satisfy the boundary conditions, we assume there is a virtual current I' in region 2.

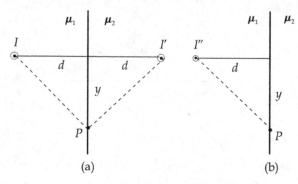

Figure 2.9 Image currents near a plane boundary.

We designate case (b) to be the situation when the observation point is inside the iron in region 2. In this case, there is no conduction current in the observation region. We assume the resultant field is due to a virtual current I'' in region 1. By symmetry, the original current and the two image currents lie on a line perpendicular to the boundary. We assume that the currents all flow in the same direction and that the distances of the currents from the boundary are equal and look for a solution for the magnitude of the currents. Consider a point P along the boundary at a distance y from the line connecting the currents. In case (a), the boundary conditions are

$$H_t^{(a)} = \frac{d}{2\pi(d^2 + y^2)}\ (I - I')$$

$$B_n^{(a)} = \frac{\mu_1\, y}{2\pi(d^2 + y^2)}\ (I + I'),$$

whereas for case (b) we have

$$H_t^{(b)} = \frac{d}{2\pi(d^2 + y^2)}\ I''$$

$$B_n^{(b)} = \frac{\mu_2\, y}{2\pi(d^2 + y^2)}\ I''.$$

Both cases must give the same solution for any position y along the boundary. Thus, we obtain the two equations

$$I - I' = I''$$
$$\mu_1 I + \mu_1 I' = \mu_2 I''.$$

Solving these two equations for the unknown magnitudes of I' and I'', we find the solution [12]

$$I' = \frac{\mu_2 - \mu_1}{\mu_2 + \mu_1}\ I \qquad\qquad (2.25)$$

and

$$I'' = \frac{2\mu_1}{\mu_1 + \mu_2}\ I. \qquad\qquad (2.26)$$

In the special case (a) when region 1 is vacuum and region 2 is infinitely permeable iron, Equation 2.25 reduces to $I' = I$.

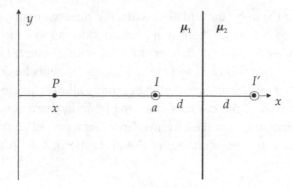

Figure 2.10 Line current near a planar iron slab.

Example 2.2:   Field enhancement in a planar iron slab

Consider a filamentary conductor a distance d away from an infinite slab of iron. We use the method of images and replace the iron slab with a virtual filament, as shown in Figure 2.10.

Let us examine the field for locations P along the line perpendicular to the boundary. Assume the observation point is at x and the filament is located at a. Then the boundary is located at $a + d$, and the image current is at $a + 2d$. The field at P is

$$B_y(x) = -\frac{\mu_0 I}{2\pi(a - x)} - \frac{\mu_0 I'}{2\pi(a + 2d - x)}.$$

Assuming $\mu_1 = \mu_0$ and $\mu_2 = \mu_r \mu_0$ and using Equation 2.25, we find that

$$B_y(x) = -\frac{\mu_0 I}{2\pi}\left[\frac{1}{a - x} + \left(\frac{\mu_r - 1}{\mu_r + 1}\right)\frac{1}{a + 2d - x}\right].$$

In the limit when $\mu_r \to \infty$ we find that $B_y(a + d) = 0$, as it should at the surface of the iron. We define the iron *enhancement factor* $E(x)$ to be the ratio of the field at x with the iron present to the field at x from the conductor by itself. In the limit $\mu_r \to \infty$, the enhancement factor is

$$E(x) = 1 + \frac{a - x}{a + 2d - x}.$$

We show the dependence of the enhancement factor on x in Figure 2.11. The enhancement is greater than 1 in the region to the left of the filament and then becomes less than 1 in the region between the filament and the iron boundary.

It is also possible to use the method of images to solve problems with more complicated arrangements of planar surfaces. For example, a line current between two parallel iron boundaries can be solved using an infinite series of image

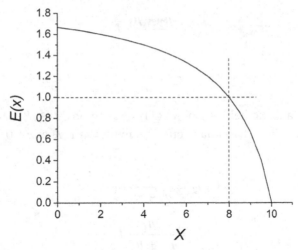

Figure 2.11 Planar iron enhancement factor for $a = 8$ and $d = 2$.

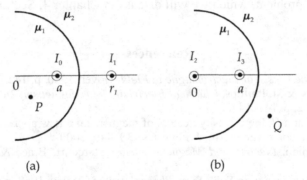

(a)   (b)

Figure 2.12 Images for a line current near a circular boundary.

currents,[13] and a line current near the corner of two perpendicular iron planes can be solved using three image currents.[5]

The method of images is also useful when considering a line current near a circular boundary surface. Consider a line current I_0 at radius a in a region with permeability μ_1 near a circular boundary of radius R of a region with permeability μ_2, as shown in Figure 2.12.

In case (a), the magnetic field at a point P inside the aperture of the magnet ($r < a$) can be written as the sum of the fields due to the conduction current I_0 at $r = a$ and the image current I_1 at $r = r_1$ inside region 2, where

[5] This is discussed in Chapter 5, Section 5.13.

$$I_1 = \frac{\mu_2 - \mu_1}{\mu_2 + \mu_1} \, I_0 \qquad\qquad (2.27)$$

$$r_1 = \frac{R^2}{a}. \qquad\qquad (2.28)$$

In case (b), the magnetic field at a point Q inside region 2 $(r > R)$ can be written as the sum of the field from two image currents in region 1, I_2 at $r = 0$ and I_3 at $r = a$, where

$$I_2 = \frac{\mu_2 - \mu_1}{\mu_2 + \mu_1} \, I_0 \qquad\qquad (2.29)$$

$$I_3 = \frac{2\mu_1}{\mu_2 + \mu_1} \, I_0. \qquad\qquad (2.30)$$

The justifications for these statements come from the solution of the corresponding boundary value problem, which we will discuss in Chapter 4, Section 4.2.

References

[1] L. Eyges, *The Classical Electromagnetic Field*, Dover, 1980, p. 142–145.
[2] W. Panofsky & M. Phillips, *Classical Electricity and Magnetism*, 2nd ed., Addison-Wesley, 1962, p. 143–144.
[3] J. Roche, *B* and *H*, the intensity vectors of magnetism: a new approach to resolving a century old controversy, Am. J. Phys. 68:438–449, 2000.
[4] D. Tomboulian, *Electric and Magnetic Fields*, Harcourt, Brace & World, 1965, p. 213–217.
[5] C. Kittel, *Introduction to Solid State Physics*, 3rd ed., Wiley, 1968, p. 432.
[6] http://magweb.us/free-bh-curves
[7] C. Rudowicz & H. Sung, Textbook treatments of the hysteresis loop for ferromagnets: survey of misconceptions and misinterpretations, *Am. J. Phys.* 71:1080–1083, 2003.
[8] G. Harnwell, *Principles of Electricity and Magnetism*, 2nd ed., McGraw-Hill, 1949, p. 403–406.
[9] Magnetically soft materials, *American Society for Metals (ASM) Handbook*, vol. 2, 1990, p. 761–781.
[10] P. Lorrain & D. Corson, *Electromagnetic Fields and Waves*, 2nd ed., Freeman, 1970, p. 405.
[11] P. Hammond, Electric and magnetic images, *Inst. Elec. Eng. Monograph* 379:306, 1960.
[12] P. Silvester, *Modern Electromagnetic Fields*, Prentice-Hall, 1968, p. 178–179.
[13] K. Binns & P. Lawrenson, *Analysis and Computation of Electric and Magnetic Field Problems*, Pergamon Press, 2nd ed., 1973, p. 38–45.

3

Potential theory

The use of potential functions represents another approach to solving problems in magnetostatics. We first treat the vector potential A. The physical significance of the vector potential has been debated for many years and it plays an important role in time-dependent phenomena.[1] However in classical magnetostatics, the potentials are usually treated as auxiliary mathematical quantities that are used to simplify the calculation of the magnetic fields. A scalar potential V_m can also be defined that satisfies Laplace's equation in current-free regions.

3.1 Vector potential

In our discussion of curl B in Chapter 1, we saw that the magnetic field could be expressed in the form given by Equation 1.21.

$$\vec{B} = \frac{\mu_0}{4\pi} \nabla \times \int \frac{\vec{J}}{R} dV. \tag{3.1}$$

We can rewrite this equation as

$$\vec{B} = \nabla \times \vec{A}, \tag{3.2}$$

where we define the *vector potential A* as

$$\vec{A} = \frac{\mu_0}{4\pi} \int \frac{\vec{J}}{R} dV. \tag{3.3}$$

This equation is valid in rectangular coordinates,[2] where the direction of A is the same as that of J.

Since B is defined as the curl of a vector by Equation 3.2, the divergence equation for B, $\nabla \cdot \vec{B} = 0$, is automatically satisfied because of the vector identity $\nabla \cdot \nabla \times \vec{V} = 0$. Using Stokes's theorem, we have

$$\oint \overrightarrow{A} \cdot \overrightarrow{dl} = \int (\nabla \times \overrightarrow{A}) \cdot \overrightarrow{dS} = \int \overrightarrow{B} \cdot \overrightarrow{dS}.$$

Thus the magnetic flux through some surface S is given by the contour integral of A around the perimeter of S.

$$\Phi_B = \oint \overrightarrow{A} \cdot \overrightarrow{dl}. \tag{3.4}$$

Any solution for the vector potential is unique, provided that the sources are confined to a finite region of space.[3]

There is an uncertainty in the defining relation for A in Equation 3.2. Any vector whose curl vanishes can be added to A without affecting the value of B. For example, the gradient of a scalar function ψ can be added because

$$\nabla \times \nabla \psi(r) = 0.$$

This freedom can be used to fix the divergence of A. Starting with Equation 3.3, the divergence of the vector potential can be written as

$$\nabla \cdot \overrightarrow{A} = \frac{\mu_0}{4\pi} \mathbb{I},$$

where

$$\mathbb{I} = \int \overrightarrow{J}(r') \nabla \cdot \left(\frac{1}{|\overrightarrow{r} - \overrightarrow{r'}|} \right) dV'.$$

Primes are used to indicate source coordinates. Note that the ∇ operator is defined in terms of the observation point (field) coordinates, while the integration and current density depend on source coordinates. Replacing the ∇ operator with one defined in terms of source coordinates, we find

$$\mathbb{I} = -\int \overrightarrow{J}(r') \nabla' \cdot \left(\frac{1}{|\overrightarrow{r} - \overrightarrow{r'}|} \right) dV'.$$

Now we can integrate using Equation B.3 and find that

$$\mathbb{I} = -\int \nabla' \cdot \left(\frac{\overrightarrow{J}\left(\overrightarrow{r'}\right)}{|\overrightarrow{r} - \overrightarrow{r'}|} \right) dV' + \int \frac{1}{|\overrightarrow{r} - \overrightarrow{r'}|} \nabla' \cdot \overrightarrow{J}\left(\overrightarrow{r'}\right) dV'.$$

The second term on the right-hand side vanishes because $\nabla' \cdot \overrightarrow{J} = 0$ from Equation 1.4. We use the divergence theorem to transform the other integral and get

$$\mathbb{I} = -\int \left(\frac{\vec{J}(\vec{r}')}{|\vec{r} - \vec{r}'|}\right) \cdot \vec{dS}.$$

We choose to evaluate this integral over a large radius sphere where all the current sources vanish. Then $\mathbb{I} = 0$ and we conclude that

$$\nabla \cdot \vec{A} = 0. \tag{3.5}$$

This result is equivalent to setting the arbitrary scalar function $\psi = 0$ and is known as the *Coulomb gauge*. Equation 3.5 represents a constraint on the components of the vector A.

Using Equation 1.24, we can relate the vector potential to the conduction current density J.

$$\nabla \times (\nabla \times \vec{A}) = \mu_0 \vec{J}, \tag{3.6}$$

Then, using the vector identity B.7, we obtain

$$\nabla(\nabla \cdot \vec{A}) - \nabla^2 \vec{A} = \mu_0 \vec{J}.$$

Using Equation 3.5, we can eliminate the gradient term and find that

$$\nabla^2 \vec{A} = -\mu_0 \vec{J}. \tag{3.7}$$

This is the vector *Poisson equation* for A. The solution is valid both inside and outside of the conductor. Equation 3.3 is a particular solution of this equation. When $J = 0$, Equation 3.7 is called the *Laplace equation*. Solutions of the Laplace equation are known as *harmonic functions*. In Cartesian coordinates, the quantity $\nabla^2 \vec{A}$ has the three scalar components $\nabla^2 A_\alpha$, where α corresponds to x, y, and z. For example,

$$\nabla^2 A_x = -\mu_0 J_x.$$

In non-Cartesian coordinate systems, the components of $\nabla^2 \vec{A}$ must be found from Equation B.7.

3.2 Vector potential in two dimensions

In cases where the current density is constant in one dimension, it is possible to develop a two-dimensional version of the theory that is significantly simpler than the general three-dimensional case. Of course this is only an approximation to the real world, but the approximation may be quite good, for example in the central region of a long magnet, far from its ends. The great power of the two-dimensional approximation will be most apparent in Chapter 5, where we develop the theory using complex analysis.

Let us consider an infinitely long current filament along the z axis. Direct integration using Equation 3.3 to $\pm\infty$ diverges. It is possible to perform the integration over a long, but finite length and then to develop an expression for A_z in a power series in ρ/L.[4] However, we will adopt another approach using the differential Equation 3.2. The field from the filament was given in Equation 1.15 and so we have

$$\nabla \times \overrightarrow{A} = \frac{\mu_0 I}{2\pi \rho}\hat{\phi}.$$

The ρ component of the curl in cylindrical coordinates gives

$$\frac{\partial A_\rho}{\partial z} - \frac{\partial A_z}{\partial \rho} = \frac{\mu_0 I}{2\pi \rho}.$$

The vector potential is constant in the z direction, so the derivative with respect to z vanishes and we have

$$-dA_z = \frac{\mu_0 I}{2\pi}\frac{d\rho}{\rho}.$$

Integrating this equation, we find

$$A_z = -\frac{\mu_0 I}{2\pi}\ln(\rho) + c.$$

If write the constant of integration c in terms of the value of A_z at some reference radius ρ_0, then

$$c = \frac{\mu_0 I}{2\pi}\ln(\rho_o)$$

and the vector potential for the infinite current filament is

$$A_z(\rho) = -\frac{\mu_0 I}{2\pi}\ln\left(\frac{\rho}{\rho_o}\right). \tag{3.8}$$

The presence of any constant terms in A_z is not important since they will be removed when taking derivatives to find B.

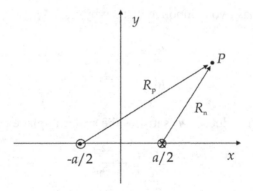

Figure 3.1 Bifilar conductors.

Example 3.1:   bifilar conductors

One method of handling the issue of the reference radius is to argue that a single filament of current is not physical because any real current must be part of a closed circuit and therefore must have a return filament somewhere with the current flowing in the opposite direction. Consider the bifilar configuration shown in Figure 3.1. Using Equation 3.8 for both the positive (p) and negative (n) filaments, we find

$$A_z = -\frac{\mu_0 I}{2\pi} \left[\ln(R_p) - \ln(R_{po}) - \ln(R_n) + \ln(R_{no}) \right],$$

where R is the distance from the current element to the observation point P. If both filaments return at the same reference position, the dependence on R_0 drops out and we find

$$A_z(x,y) = -\frac{\mu_0 I}{2\pi} \ln\left(\frac{R_p}{R_n}\right). \tag{3.9}$$

The sign of A_z depends on the relative magnitudes of R_p and R_n.

We can find the two-dimensional vector potential for current distributions by superposition of the vector potential for a filament. The vector potential for a two-dimensional current sheet is given by

$$A_z(x,y) = -\frac{\mu_0}{2\pi} \int K_z(s') \ln\left(\frac{R}{R_o}\right) ds', \tag{3.10}$$

where $K_z = dI/ds$ is the sheet current density. The vector potential for a block conductor with finite cross-section is

$$A_z(x,y) = -\frac{\mu_0}{2\pi} \iint J_z(x',y') \ln\left(\frac{R}{R_o}\right) dx' dy'. \tag{3.11}$$

The magnetic field has two components given by

$$B_x = \frac{\partial A_z}{\partial y}$$
$$B_y = -\frac{\partial A_z}{\partial x}$$

(3.12)

and, except inside a conductor, A_z satisfies the scalar Laplace equation

$$\nabla^2 A_z = \frac{\partial^2 A_z}{\partial x^2} + \frac{\partial^2 A_z}{\partial y^2} = 0.$$

(3.13)

3.3 Boundary conditions on A

The boundary conditions on the vector potential A can be determined from the boundary conditions on the magnetic field. Consider a boundary surface S located at the intersection of two regions of space. We know from Equation 2.22 that the normal component of B must be conserved across S. Thus the magnetic flux crossing S is conserved and from Equation 3.4

$$\oint \vec{A} \cdot \vec{dl} = \Phi_B,$$

we see that the line integral of A around the perimeter of S must also be conserved. Therefore the tangential component of A must be conserved on the boundary.

$$\vec{A}_t^{(1)} = \vec{A}_t^{(2)}.$$

(3.14)

The boundary condition on the tangential component of H given in Equation 2.23 can be written in the form

$$\frac{1}{\mu^{(2)}} (\nabla \times \vec{A})_t^{(2)} - \frac{1}{\mu^{(1)}} (\nabla \times \vec{A})_t^{(1)} = K,$$

(3.15)

where K is the surface current on S, if applicable. These two vector relations provide four constraints on A at the boundary.[3]

The boundary conditions can be considerably simpler in two dimensions. Assume, for example, that the problem is uniform in the z direction and that we have a boundary between two regions, as shown in Figure 3.2. All the current is along z, so A only has the component A_z. In this case, the boundary surface is parallel to the x-z plane and the tangential component of A is along z. From Equation 3.14, we have

$$A_z^{(1)} = A_z^{(2)},$$

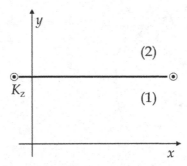

Figure 3.2 Boundary in two dimensions.

so the vector potential is continuous across the boundary. Since

$$\nabla \times \vec{A} = \hat{x}\partial_y A_z - \hat{y}\,\partial_x A_z,$$

the tangential component of the curl is along x. Thus Equation 3.15 gives

$$\frac{1}{\mu^{(2)}}\,\partial_y A_z^{(2)} - \frac{1}{\mu^{(1)}}\,\partial_y A_z^{(1)} = K_z.$$

In this case, we have two constraints that must be satisfied at the boundary.

3.4 Vector potential for a localized current distribution

Consider a localized distribution of current, as shown in Figure 3.3. We pick some origin O inside the distribution and examine the potential at a field location P. Since the distance between the source point at r' and the field point at r is

$$R = \{r^2 + r'^2 - 2rr'\cos\theta\}^{1/2},$$

we can express the inverse distance as

$$\frac{1}{R} \simeq \frac{1}{r} + \frac{\vec{r}\cdot\vec{r}'}{r^3} + \cdots.$$

Then the first two terms in the multipole expansion for A are

$$\vec{A}(\vec{r}) \simeq \frac{\mu_0}{4\pi}\left[\frac{1}{r}\int \vec{J}(\vec{r}')\,dV' + \frac{\vec{r}}{r^3}\cdot\int \vec{r}'\vec{J}(\vec{r}')\,dV' + \cdots\right]. \qquad (3.16)$$

The first integral vanishes because the current distribution consists of closed loops. In the second integral, J corresponds to one of the components of A,

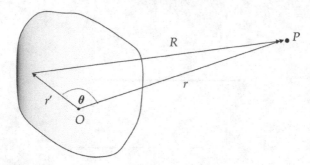

Figure 3.3 Localized current distribution.

which we specify with the index i, while r' is part of the scalar product with r, which we specify with the index j. Thus the integral has the form

$$\mathbb{I} = \int r'_j J'_i \, dV'.$$

Using the vector identity B.3, we have

$$\nabla' \cdot (r'_i \overrightarrow{J'}) = r'_i \nabla' \cdot \overrightarrow{J'} + \overrightarrow{J'} \cdot \nabla' r'_i.$$

The first term vanishes because of Equation 1.4 and in the second term, we have $\nabla' r'_i = \hat{i}$. Thus

$$J'_i = \nabla' \cdot (r'_i \overrightarrow{J'})$$

and so we have

$$\mathbb{I} = \int \nabla' \cdot (r'_i \overrightarrow{J'}) r'_j \, dV'.$$

We can do the integral by parts with

$$u = r'_j$$
$$dv = \nabla' \cdot \left(r'_i \overrightarrow{J'} \right).$$

This gives

$$\mathbb{I} = \int r'_i r'_j \overrightarrow{J'} \, dS' - \int r'_i \overrightarrow{J'} \cdot \nabla' r'_j \, dV'.$$

The first term vanishes for a surface outside the charge distribution. In the second term, the gradient in the integrand vanishes except for the j component. Thus

$$\mathbb{I} = -\int r_i' J_j'\, dV'.$$

Comparing with Equation 3.16, this implies that

$$\int (\vec{r} \cdot \vec{r}')\, \vec{J}\, dV' = -\int (\vec{r} \cdot \vec{J})\, \vec{r}'\, dV'. \tag{3.17}$$

Now consider the triple vector product from Equation B.1.

$$\vec{r} \times (\vec{r}' \times \vec{J}) = \vec{r}'(\vec{r} \cdot \vec{J}) - \vec{J}(\vec{r} \cdot \vec{r}').$$

Substituting Equation 3.17, we have

$$2\int (\vec{r} \cdot \vec{r}')\vec{J}\, dV' = -\int \vec{r} \times (\vec{r}' \times \vec{J})\, dV'. \tag{3.18}$$

Substituting this back into Equation 3.16, we find that the vector potential is

$$\vec{A}(\vec{r}) \simeq -\frac{1}{2} \frac{\mu_0}{4\pi} \frac{\vec{r}}{r^3} \times \int \vec{r}' \times \vec{J}\, dV'.$$

We define the magnetic moment of the current distribution as

$$\vec{m} = \frac{1}{2} \int \vec{r}' \times \vec{J}(\vec{r}')\, dV'. \tag{3.19}$$

Then the vector potential of the current distribution can be written as [5]

$$\vec{A}(\vec{r}) = \frac{\mu_0}{4\pi} \frac{\vec{m} \times \vec{r}}{r^3}. \tag{3.20}$$

Thus we find that the elementary form of magnetic matter is a magnetic dipole.

In spherical coordinates, the vector potential for a magnetic dipole is directed in the azimuthal direction. We can find the magnetic field from the dipole by taking the curl of A.

$$\begin{aligned} B_r &= \frac{1}{r \sin \theta} \partial_\theta (A_\phi \sin \theta) \\ &= \frac{\mu_0}{4\pi} \frac{2m}{r^3} \cos \theta \end{aligned} \tag{3.21}$$

and

$$\begin{aligned} B_\theta &= -\frac{1}{r} \partial_r (r A_\phi) \\ &= \frac{\mu_0}{4\pi} \frac{m}{r^3} \sin \theta. \end{aligned} \tag{3.22}$$

The remaining component $B_\phi = 0$. We see that the field of a magnetic dipole falls off like $1/r^3$.

We can relate this definition of the magnetic moment, Equation 3.19, with our discussion in Chapter 1 of the magnetic moment of a planar current loop. We let

$$\vec{J}\, dV' \rightarrow I\, \vec{dl'}.$$

Then Equation 3.19 gives

$$\vec{m} = \tfrac{1}{2} I \oint \vec{r}\,' \times \vec{dl'}.$$

The magnitude of the quantity $\tfrac{1}{2}\vec{r}\,' \times \vec{dl'}$ is the area of a triangular region inside the current loop. The closed integral then gives the total area A enclosed by the loop. Thus

$$\vec{m} = I\, A\, \hat{n},$$

which agrees with the result in Equation 1.10.

Now consider a volume of magnetic material that contains many magnetic dipoles. The contribution to the vector potential from one small part of the overall volume at r' can be written in terms of the magnetization vector M as

$$\vec{dA} = \frac{\mu_0}{4\pi} \frac{\vec{M'} \times \vec{R}}{R^3}\, dV'.$$

Using the relation

$$\nabla\left(\frac{1}{R}\right) = -\frac{1}{R^2}\hat{r} = -\nabla'\left(\frac{1}{R}\right),$$

the vector potential for the whole volume is

$$\vec{A} = \frac{\mu_0}{4\pi} \int \vec{M'} \times \nabla'\left(\frac{1}{R}\right) dV'.$$

Using the vector identity B.6, we can write A as the two terms

$$\vec{A} = -\frac{\mu_0}{4\pi} \int \nabla' \times \left(\frac{\vec{M'}}{R}\right) dV' + \frac{\mu_0}{4\pi} \int \frac{\nabla' \times \vec{M'}}{R}\, dV'.$$

Then using vector identity

$$\int \nabla \times \vec{W}\, dV = -\int \vec{W} \times \hat{n}\, dS$$

in the first term and dropping the primes, we get

$$\vec{A} = \frac{\mu_0}{4\pi} \int \frac{\vec{M} \times \hat{n}}{R} dS + \frac{\mu_0}{4\pi} \int \frac{\nabla \times \vec{M}}{R} dV. \tag{3.23}$$

We have seen previously from Equation 2.2 that $\vec{K}_m = \vec{M} \times \hat{n}$ is the surface current density and from Equation 2.5 that $\vec{J} = \nabla \times \vec{M}$ is the volume current density.

3.5 Force on a localized current distribution

The force on a localized current in an applied magnetic field is

$$\vec{F} = \int \vec{J} \times \vec{B} \, dV'.$$

If the field is non-uniform, each component of B can be expanded in a Taylor's series.

$$B_i(\vec{r}) \simeq B_i(0) + \vec{r} \cdot \nabla B_i(0) + \cdots.$$

Then

$$\vec{F} \simeq - \vec{B}(0) \times \int \vec{J}(\vec{r}') dV' + \int \vec{J}(\vec{r}') \times \left[(\vec{r}' \cdot \nabla) \vec{B}(0) \right] dV'.$$

The first integral vanishes since J consists of closed loops. Using Equation B.2, we have

$$\nabla(\vec{r}' \cdot \vec{B}) = \vec{r}' \times (\nabla \times \vec{B}) + \vec{B} \times (\nabla \times \vec{r}') + (\vec{B} \cdot \nabla) \vec{r}' + (\vec{r}' \cdot \nabla) \vec{B}.$$

The first term on the right-hand side vanishes outside the current distribution, while the second and third terms vanish because the ∇ operator refers to field coordinates. Thus the force can be written as

$$\vec{F} = \int \vec{J} \times \nabla(\vec{r}' \cdot \vec{B}) dV'.$$

From Equation B.6, we have

$$\nabla \times [\vec{J}(\vec{r}' \cdot \vec{B})] = (\vec{r}' \cdot \vec{B}) \nabla \times \vec{J} - \vec{J} \times \nabla(\vec{r}' \cdot \vec{B}).$$

The first term on the right side vanishes because ∇ operates on field coordinates, so

$$\vec{F} = -\nabla \times \int \vec{J'}(\vec{r'} \cdot \vec{B}) dV'.$$

Using the relation in Equation 3.18 with r replaced by B, we have

$$\int (\vec{B} \cdot \vec{r'}) \vec{J'} \, dV' = -\tfrac{1}{2} \vec{B} \times \int \vec{r'} \times \vec{J'} dV'.$$

Thus

$$\vec{F} = -\nabla \times \int \left[-\tfrac{1}{2} \vec{B} \times (\vec{r'} \times \vec{J'}) \right] dV'.$$

Using Equation 3.19, we can write this in terms of the magnetic moment as

$$\vec{F} = \nabla \times (\vec{B} \times \vec{m'}).$$

Finally, using Equation B.9

$$\vec{F} = \vec{B} \, (\nabla \cdot \vec{m'}) - \vec{m'} \, (\nabla \cdot \vec{B}) + (\vec{m'} \cdot \nabla) \, \vec{B} - (\vec{B} \cdot \nabla) \vec{m'}.$$

The first and fourth terms vanish because m is independent of the field coordinates. Thus dropping the primes, we find that the force on a magnetic dipole in an inhomogeneous magnetic field is given by

$$\vec{F} = (\vec{m} \cdot \nabla) \, \vec{B}. \tag{3.24}$$

For a continuous magnetization distribution, the force is given by [6]

$$\vec{F} = \int (\vec{M} \cdot \nabla) \vec{B} \, dV. \tag{3.25}$$

3.6 Magnetic scalar potential

In regions of space where there are no conduction currents present, the magneto-static field equations become

$$\nabla \times \vec{H} = 0$$
$$\nabla \cdot \vec{B} = 0.$$

From the curl equation, we know that the magnetic field can be expressed as the gradient of a scalar function

$$\vec{H} = -\nabla V_m, \tag{3.26}$$

which we call the *magnetic scalar potential* V_m. Multiplying this equation by μ_0 and substituting the expression for B into the divergence equation, we find that

$$\nabla^2 V_m = 0. \tag{3.27}$$

Thus V_m satisfies the Laplace equation in regions where the conduction current $J = 0$.

The boundary condition on the transverse component of the field H_t in Equation 2.23 gives

$$H_t^{(2)} - H_t^{(1)} = K$$
$$-\partial_t V_m^{(2)} + \partial_t V_m^{(1)} = K. \tag{3.28}$$

In the case where no surface current is present, the gradients in the two regions must be the same. This implies that the scalar potentials in the two regions can only differ at most by a constant factor c.

$$V_m^{(2)} = V_m^{(1)} + c$$

From the boundary condition on B_n in Equation 2.22, we have

$$B_n^{(2)} = B_n^{(1)}$$
$$\mu^{(2)}\, \partial_n V_m^{(2)} = \mu^{(1)}\, \partial_n V_m^{(1)}. \tag{3.29}$$

Thus the product of the permeability and the normal derivative of V_m is continuous across the boundary.

Consider a field point P nearby a current loop, as shown in Figure 3.4. Assign a normal vector n to the loop according to the right hand rule. Now displace the loop by the vector du. The cross product of dl with du gives the area of the shaded area in the figure. In this case, the solid angle at P subtended by the differential area dS changes by an amount

$$d\Omega = \frac{d\vec{S} \cdot \hat{r}}{r^2} = \frac{(\vec{du} \times \vec{dl}) \cdot \hat{r}}{r^2} = \frac{(\vec{dl} \times \hat{r}) \cdot \vec{du}}{r^2}.$$

The last equation makes use of the fact that the terms in the triple vector product permute. To get the total change in solid angle, we sum over all the parts of the loop.

$$d\Omega = \vec{du} \cdot \oint \frac{\vec{dl} \times \vec{r}}{r^3}$$
$$= \vec{du} \cdot \nabla\Omega$$

Comparing these two equations, we find that the gradient of the solid angle is given by

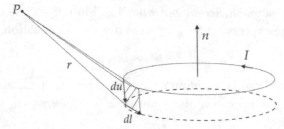

Figure 3.4 Solid angle of a displaced current loop.

$$\nabla\Omega = \oint \frac{\overrightarrow{dl} \times \overrightarrow{r}}{r^3}. \tag{3.30}$$

Now we want to relate this to the scalar potential

$$\nabla V_m = -\frac{\overrightarrow{B}}{\mu}.$$

Expressing B using the Biot-Savart law and using Equation 3.30, we find

$$\nabla V_m = -\frac{I}{4\pi}\oint \frac{\overrightarrow{dl} \times \overrightarrow{r}}{r^3}$$

$$= -\frac{I}{4\pi}\nabla\Omega.$$

The gradients on the two sides of the equation are proportional to each other, so V_m must depend linearly on Ω.

$$V_m = -\frac{I}{4\pi}\Omega. \tag{3.31}$$

This indicates that the magnetic scalar potential is directly related to the solid angle that a current loop subtends at a field observation point. We can ignore any constant term since it drops out when calculating the field. Note that Equation 3.31 implies that the scalar potential is not a single-valued function. If P moves from just in front of the loop to just behind it, the solid angle changes from 2π to -2π since $\hat{n}\cdot\hat{r}$ changes sign.

3.7 Scalar potential for a magnetic body

In analogy with the potential due to dielectric polarization in electrostatics,[7] the scalar magnetic potential associated with the magnetization vector is given by

$$V_m = \frac{1}{4\pi} \int \frac{\vec{M} \cdot \vec{R}}{R^3} dV, \tag{3.32}$$

which we can express as

$$V_m = \frac{1}{4\pi} \int \vec{M'} \cdot \nabla' \left(\frac{1}{R}\right) dV'.$$

Using the vector identity B.3, we can write this as

$$V_m = \frac{1}{4\pi} \left[\int \nabla' \cdot \frac{\vec{M'}}{R} dV' - \int \frac{\nabla' \cdot \vec{M'}}{R} dV' \right].$$

Using the divergence theorem for the first term and dropping the primes gives

$$V_m = \frac{1}{4\pi} \left[\int \frac{\vec{M} \cdot \hat{n}}{R} dS - \int \frac{\nabla \cdot \vec{M}}{R} dV \right]. \tag{3.33}$$

We can identify the first term on the right-hand side as coming from fictitious magnetic charges on the surface of the magnetized body with the surface current density[8]

$$K_m = \vec{M} \cdot \hat{n} \tag{3.34}$$

and the second term can be described as due to the volume charge density

$$\rho_m = -\nabla \cdot \vec{M}. \tag{3.35}$$

This shows that for the purpose of finding the magnetic field outside a magnetized body, we can replace the body with equivalent magnetic surface and volume charges.

3.8 General solutions to the Laplace equation

We have found that single components of the vector potential and the scalar potential both satisfy the scalar Laplace equation $\nabla^2 F = 0$ outside conductor regions. In general, this is a three-dimensional partial differential equation. Solutions of the scalar Laplace equation depend on the coordinate system that is used.

A common technique for solving the Laplace equation is to use the method of *separation of variables*. This method assumes that the solution is the product of three terms, each of which only depends on one of the coordinates. As a result, the partial differential equation in three variables is converted into three ordinary

Potential theory

differential equations, each of which depends on a single variable. The method is most useful when a boundary surface in the problem lies along one of the coordinate directions. The Laplace equation is known to be separable in eleven coordinate systems.[9] We summarize results here for the three most common systems.

Rectangular coordinates [10]

The Laplace equation in rectangular coordinates is

$$\frac{\partial^2 F}{\partial x^2} + \frac{\partial^2 F}{\partial y^2} + \frac{\partial^2 F}{\partial z^2} = 0. \tag{3.36}$$

We assume the solutions can be written in the form

$$F(x,y,z) = X(x)Y(y)Z(z).$$

When this is substituted into Equation 3.36, we obtain

$$\frac{1}{X}\frac{d^2 X}{dx^2} + \frac{1}{Y}\frac{d^2 Y}{dy^2} + \frac{1}{Z}\frac{d^2 Z}{dz^2} = 0.$$

This equation can only be valid for all values of x, y, or z if each of the three terms is equal to a constant. Thus we have

$$\frac{1}{X}\frac{d^2 X}{dx^2} = a^2$$

$$\frac{1}{Y}\frac{d^2 Y}{dy^2} = b^2 \tag{3.37}$$

$$\frac{1}{Z}\frac{d^2 Z}{dz^2} = c^2,$$

where the three constants have to satisfy the constraint

$$a^2 + b^2 + c^2 = 0. \tag{3.38}$$

In order to satisfy this equation, the set of constants $\{a, b, c\}$ must contain both real and imaginary members. The choice of which constants are real and which are imaginary depends on the specific conditions for the problem under consideration.

As an example, let us suppose that the constants a and b are imaginary and c is real. Then we can define a new set of real constants $\{\alpha, \beta, \gamma\}$ such that

$$a^2 = (i\alpha)^2 = -\alpha^2$$
$$b^2 = (i\beta)^2 = -\beta^2$$
$$c^2 = \gamma^2.$$

In the general case, there will be a set of constants that satisfy Equation 3.37.

$$\alpha_n = \{\alpha_1, \alpha_2, \ldots\}$$
$$\beta_m = \{\beta_1, \beta_2, \ldots\}$$

The constraint Equation 3.38 then becomes

$$-\alpha_n^2 - \beta_m^2 + \gamma_{nm}^2 = 0$$

and the separated differential equations are

$$\frac{1}{X_n} \frac{d^2 X_n}{dx^2} = -\alpha_n^2$$

$$\frac{1}{Y_m} \frac{d^2 Y_m}{dy^2} = -\beta_m^2$$

$$\frac{1}{Z_{nm}} \frac{d^2 Z_{nm}}{dz^2} = \gamma_{nm}^2.$$

Then the general solution in Cartesian coordinates has the form

$$F(x, y, z) = \sum_{n,m=1}^{\infty} \left[C_n e^{i\alpha_n x} + D_n e^{-i\alpha_n x} \right] \left[E_m e^{i\beta_m y} + F_m e^{-i\beta_m y} \right] \left[G_{nm} e^{\gamma_{nm} z} + H_{nm} e^{-\gamma_{nm} z} \right]$$
$$+ I_0 + I_1 x + I_2 y + I_3 z.$$

The oscillatory terms could also be written in terms of sines and cosines and the nonoscillatory terms could be written using hyperbolic sines and cosines. The last four terms are also a solution of the Laplace equation that allow for continuity of the potential and the presence of external fields.

Cylindrical coordinates [11]

The Laplace equation in cylindrical coordinates is

$$\frac{1}{\rho} \partial_\rho (\rho \partial_\rho F) + \frac{1}{\rho^2} \partial_\phi^2 F + \partial_z^2 F = 0. \tag{3.39}$$

Using the method of separation of variables, we assume that

$$F(\rho, \phi, z) = R(\rho) \Phi(\phi) Z(z).$$

This leads to the three ordinary differential equations

$$\rho \frac{d}{d\rho} \left(\rho \frac{dR}{d\rho} \right) + (k^2 \rho^2 - n^2) R = 0 \tag{3.40}$$

$$\frac{d^2\Phi}{d\phi^2} + n^2\Phi = 0 \tag{3.41}$$

$$\frac{d^2Z}{dz^2} - k^2 Z = 0, \tag{3.42}$$

where k and n are constants that may be real or imaginary.

In order for the azimuthal dependence to be single-valued, the parameter n in Equation 3.41 must be a real integer. Solutions for the function Φ have the general form

$$\Phi_n(\phi) = C_n\cos(n\phi) + D_n\sin(n\phi) \tag{3.43}$$

when $n \neq 0$ and

$$\Phi_0(\phi) = C_0\,\phi + D_0 \tag{3.44}$$

when $n = 0$.

The equations for R and Z have different solutions, depending on the values for n and k.

(1) real $k \neq 0$

The general form of the z dependence in Equation 3.42 is

$$Z_k(z) = E_k e^{kz} + F_k e^{-kz}.$$

The solution for the radial dependence in Equation 3.40 has the form

$$R_n(k\rho) = G_n J_n(k\rho) + H_n N_n(k\rho),$$

where J_n and N_n are integer Bessel functions of the first and second kind.[12]

(2) imaginary $k \neq 0$

If $k = i\,\kappa$ where κ is real, then the z solution is oscillatory.

$$Z_k(z) = E_k e^{i\kappa z} + F_k e^{-i\kappa z}.$$

In this case, Equation 3.40 becomes

$$\rho\frac{d}{d\rho}\left(\rho\frac{dR}{d\rho}\right) - (\kappa^2\rho^2 + n^2)R = 0 \tag{3.45}$$

and the radial solution is

$$R_n(\kappa\rho) = G_n I_n(\kappa\rho) + H_n K_n(\kappa\rho),$$

where I_n and K_n are modified Bessel functions.[13]

(3) $k = 0$

The z solution has the form

$$Z_0(z) = E_0 z + F_0.$$

The general radial solution for $n \neq 0$ is

$$R_n(\rho) = G_n \rho^n + H_n \rho^{-n}. \tag{3.46}$$

If $n = 0$, the radial solution is

$$R_0(\rho) = G_0 \ln r + H_0. \tag{3.47}$$

The general form of the solution of the Laplace equation in cylindrical coordinates can then be written in the form

$$F(\rho, \phi, z) = \sum_{k,n} C_{kn} \, R_n \, (k\rho) \Phi_n(\phi) Z_k(z).$$

Some additional information concerning Bessel functions can be found in Appendix C.

Spherical coordinates [14]

The Laplace equation in spherical coordinates is

$$\frac{1}{r^2} \partial_r(r^2 \partial_r F) + \frac{1}{r^2 \sin \theta} \partial_\theta(\sin \theta \, \partial_\theta F) + \frac{1}{r^2 \sin^2 \theta} \partial_\phi^2 F = 0. \tag{3.48}$$

The radial and angular parts of this equation can be separated first in the form

$$F(r, \theta, \phi) = R(r)Y(\theta, \phi).$$

This leads to the radial equation

$$\frac{d}{dr}\left(r^2 \frac{dR}{dr}\right) - n(n+1)R = 0, \tag{3.49}$$

which has a general solution of the form

$$R_n(r) = G_n r^n + H_n r^{-n}.$$

The angular equation is

$$\frac{1}{\sin \theta} \partial_\theta(\sin \theta \, \partial_\theta Y) + \frac{1}{\sin^2 \theta} \partial_\phi^2 Y + n(n+1) \, Y = 0. \tag{3.50}$$

The solutions for this equation are known as *spherical harmonics*.[15] The constant n must be an integer to avoid singularities in $Y(\theta, \phi)$ at $\theta = 0$ and $\theta = \pi$. The two angle coordinates can in turn be separated as

$$Y(\theta, \phi) = \Theta(\theta)\Phi(\phi).$$

This leads to the two ordinary differential equations

$$\frac{d}{dx}\left[(1 - x^2)\frac{d\Theta}{dx}\right] + \left[n(n+1) - \frac{m^2}{1 - x^2}\right]\Theta = 0 \qquad (3.51)$$

and

$$\frac{d^2\Phi}{d\phi^2} + m^2\Phi = 0, \qquad (3.52)$$

where $x = \cos\theta$. In order for the azimuthal dependence to be single-valued in Equation 3.52, m must be an integer, and Φ has the solution

$$\Phi_m(\phi) = C_m\cos(m\phi) + D_m\sin(m\phi)$$

when $m \neq 0$ and

$$\Phi_0(\phi) = C_0\phi + D_0$$

when $m = 0$. The solution of Equation 3.51 has the form

$$\Theta_n^m(\theta) = E_{mn}P_n^m(\cos\theta) + F_{mn}Q_n^m(\cos\theta),$$

where P_n^m and Q_n^m are *associated Legendre functions* of the first and second kind.[16] In problems with azimuthal symmetry, we have $m = 0$ and the associated Legendre functions P_n^m reduce to the ordinary Legendre polynomials.

$$P_n^0(\cos\theta) = P_n(\cos\theta).$$

The general form of the solution of the Laplace equation in spherical coordinates can then be written in the form

$$F(r, \theta, \phi) = \sum_{n,m} C_{nm}R_n(r)\Theta_n^m(\theta)\Phi_m(\phi).$$

Some additional information concerning Legendre functions is given in Appendix D.

3.9 Boundary value problems

Unique solutions for boundary value problems for the Laplace equation require that either the potential F or its normal derivative be specified on the boundary.[17] Problems where F is specified on the boundary are known as *Dirichlet* boundary value problems, while problems where $\partial F / \partial n$ are specified are known as *Neumann* boundary value problems. Both of these types of problem give unique solutions for the Laplace equation. Solutions do not exist when both F and $\partial F / \partial n$ are arbitrarily specified because the derivatives of F have to be constrained to satisfy the Laplace equation.

The solution of boundary value problems begins by separating the problem space into regions with unique values for the current density and permeability. For each region, a potential function is written in the most general form possible. This introduces a set of unknown coefficients in the potential functions. Constraints on these coefficients are determined by demanding that the potential functions satisfy the boundary conditions at all the interfaces between different regions.

Fourier analysis is particularly useful in problems involving rectangular conductors in a space with rectangular boundaries.[18] If the boundaries are infinitely-permeable iron surfaces, they can be replaced with a set of image currents. Then the current density can be expressed as a Fourier series and the fields can be determined from the solution to a boundary value problem.

Example 3.2: rectangular conductor in an infinite slot
We first consider an example using the vector potential. Assume we have a rectangular-shaped conductor near the bottom of an infinitely deep slot with infinitely permeable walls, as shown in Figure 3.5. The current flows in the z direction. The current density in the conductor is given by

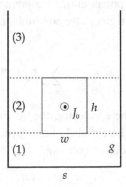

Figure 3.5 Rectangular conductor in a slot.

Potential theory

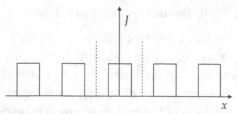

Figure 3.6 Periodic current distribution.

$$J(x) = \begin{cases} 0 & -s/2 \le x \le -w/2 \\ J_0 & -w/2 \le x \le w/2 \\ 0 & w/2 \le x \le s/2. \end{cases}$$

We replace the parallel side walls of the slot with an infinite set of image conductors, whose current density is shown in Figure 3.6. The Fourier series representing the current distribution is

$$J(x) = J_0 \frac{w}{s} + \frac{2J_0}{\pi} \sum_{n=1}^{\infty} \frac{1}{n} \sin\left(\frac{nkw}{2}\right) \cos(nkx), \qquad (3.53)$$

where $k = 2\pi/s$.

Divide the problem space vertically into three regions, as shown in Figure 3.5. In region 1, there are no currents so the vector potential A_1 satisfies the Laplace equation. To satisfy the boundary conditions, we know that the x dependence has to correspond with the x dependence of $J(x)$. Thus we have

$$A_1 = \sum_{n=1}^{\infty} (C_n e^{nky} + D_n e^{-nky}) \cos(nkx),$$

where C_n and D_n are unknown coefficients. The solution A_3 for region 3 also has to satisfy the Laplace equation. Since region 3 extends to infinite values of y, the term proportional to e^{nky} must vanish. Far from the conductor, the field must be uniform along the x direction, so the potential must contain a term proportional to y. It must also contain a constant term to guarantee continuity of A. Thus the general form of the potential in region 3 is

$$A_3 = E_0 + E_1 y + \sum_{n=1}^{\infty} F_n e^{-nky} \cos(nkx).$$

Since region 2 contains the conductor, the vector potential A_2 has to satisfy the Poisson equation. The total potential A_2 has a general (or homogeneous) part plus a particular solution to the Poisson equation.

$$A_2 = A_{2h} + A_{2p}.$$

To match the potential at the boundary with region 3, the general part of the potential has to include a constant term and a term linear in y. Thus we have

$$A_{2h} = G_0 + G_1 y + \sum_{n=1}^{\infty} (H_n e^{nky} + M_n e^{-nky}) \cos(nkx).$$

Since the current density J has a constant term and a periodic term, we look for a particular potential of the form

$$A_{2p} = G_{2p} \, y^2 + \sum_{n=1}^{\infty} L_n \cos(nkx).$$

Substitute this expression into the Poisson equation, together with Equation 3.53 for the current density. Since the expansion makes use of orthogonal functions, we can equate the constant term and each term in the series independently. We find the coefficients for the particular solution are

$$G_{2p} = -\frac{\mu_0 J_0 w}{2s}$$

and

$$L_n = \frac{2\mu_0 J_0}{\pi n^3 k^2} \sin\left(\frac{nkw}{2}\right).$$

The three equations for A contain a total of nine unknown coefficients. We find the values of these coefficients by imposing the boundary conditions.

Case 1: $y = 0$.

The bottom boundary is an infinite permeability surface, so $B = B_y$ must be perpendicular to this surface. Thus

$$B_x = \frac{\partial A_1}{\partial y} = 0,$$

which gives

$$C_n - D_n = 0. \tag{3.54}$$

Case 2: $y = g$

The nonperiodic and periodic parts of the equations for the continuity of A and $\partial_y A$ give the four equations

$$0 = G_0 + G_1 \, g - \frac{\mu_0 J_0 w}{2s} \, g^2 \tag{3.55}$$

$$C_n e^{nkg} + D_n e^{-nkg} = H_n e^{nkg} + M_n e^{-nkg} + \frac{2\mu_0 J_0}{\pi n^3 k^2} \sin\left(\frac{nkw}{2}\right) \qquad (3.56)$$

$$0 = G_1 - \frac{\mu_0 J_0 w}{s} g \qquad (3.57)$$

$$C_n e^{nkg} - D_n e^{-nkg} = H_n e^{nkg} - M_n e^{-nkg}. \qquad (3.58)$$

Case 3: $y = g + h$

The nonperiodic and periodic parts of the equations for the continuity of A and $\partial_y A$ give the four equations

$$G_0 + G_1(g+h) - \frac{\mu_0 J_0 w}{2s}(g+h)^2 = E_0 + E_1(g+h) \qquad (3.59)$$

$$H_n e^{nk(g+h)} + M_n e^{-nk(g+h)} + \frac{2\mu_0 J_0}{\pi n^3 k^2} \sin\left(\frac{nkw}{2}\right) = F_n e^{-nk(g+h)} \qquad (3.60)$$

$$G_1 - \frac{\mu_0 J_0 w}{s}(g+h) = E_1 \qquad (3.61)$$

$$H_n e^{nk(g+h)} - M_n e^{-nk(g+h)} = -F_n e^{-nk(g+h)}. \qquad (3.62)$$

Case 4: $y \to \infty$

At large y, far from the conductor, the field must be uniform, so we have

$$B_x = \frac{\partial A_3}{\partial y} = E_1.$$

From far above, the conductor looks like an infinite current sheet with current density $J_0 h$, whose strength is reduced by the filling factor w/s and is enhanced by a factor 2 due to the presence of the bottom permeable surface. Then using Equation 1.17, we find

$$E_1 = B_x = -\tfrac{1}{2} \mu_0 J_0 h \frac{w}{s} 2$$
$$= -\frac{\mu_0 J_0 h w}{s}. \qquad (3.63)$$

Equations 3.54–3.63 give ten constraints on the nine unknown coefficients. However, Equation 3.61 is redundant since it is equivalent to Equations 3.57 and 3.63. Thus we have nine equations in nine unknowns. After solving this system of equations, the resulting vector potentials are:

$$A_1 = \alpha \sum_{n=1}^{\infty} (e^{nky} + e^{-nky})\cos(nkx), \qquad (3.64)$$

where

$$\alpha = -\frac{\mu_0 J_0}{\pi n^3 k^2} \sin\left(\frac{nkw}{2}\right) e^{-nkg} \left(e^{-nkh} - 1\right); \tag{3.65}$$

$$A_2 = -\frac{\mu_0 J_0 w}{2s} (y - g)^2 + \sum_{n=1}^{\infty} (\beta_1 e^{nky} + \beta_2 e^{-nky} + \beta_3) \cos(nkx), \tag{3.66}$$

where

$$\beta_1 = -\frac{\mu_0 J_0}{\pi n^3 k^2} \sin\left(\frac{nkw}{2}\right) e^{-nk(g+h)} \tag{3.67}$$

$$\beta_2 = \frac{\mu_0 J_0}{\pi n^3 k^2} \sin\left(\frac{nkw}{2}\right) \left(-e^{-nk(g+h)} - e^{nkg} + e^{-nkg}\right) \tag{3.68}$$

$$\beta_3 = 2\mu_0 J_0 \sin\left(\frac{nkw}{2}\right); \tag{3.69}$$

$$A_3 = \frac{\mu_0 J_0 hw(2g+h)}{2s} - \frac{\mu_0 J_0 hw}{s} y + \sum_{n=1}^{\infty} \gamma e^{-nky} \cos(nkx), \tag{3.70}$$

where

$$\gamma = -\frac{\mu_0 J_0}{\pi n^3 k^2} \sin\left(\frac{nkw}{2}\right) \left(-e^{nk(g+h)} + e^{-nk(g+h)} + e^{nkg} - e^{-nkg}\right). \tag{3.71}$$

The series for the vector potential converge rapidly because of the n^3 factor in the denominators of the coefficients.

We find B by taking the curl of the vector potential. The resulting field in the slot is shown in Figure 3.7. The dotted lines show the location of the conductor.

Example 3.3: permeable sphere in external magnetic field

We next consider an example of using the scalar potential. Assume we have a sphere of some magnetic material located in an external magnetic field, as shown in Figure 3.8. We choose the z coordinate of a spherical coordinate system to lie along the direction of the external field $B_0 = \mu_0 H_0$. The problem has azimuthal symmetry, so the results cannot depend on ϕ. The magnetic scalar potential for the external field can be written as

$$V_0 = -H_0 z = -H_0 r \cos\theta$$
$$= -H_0 r P_1(\cos\theta),$$

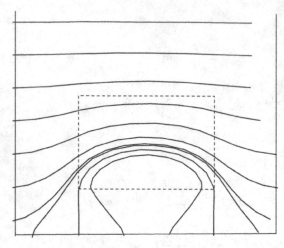

Figure 3.7 Magnetic flux density inside the slot.

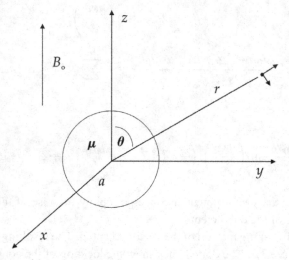

Figure 3.8 Permeable sphere in an external magnetic field.

where P_1 is a Legendre polynomial. The potential for the problem including the sphere must remain finite as $r \rightarrow \infty$. Therefore the potential outside the sphere is given by

$$V_m^{ext} = V_0 + \sum_{n=0}^{\infty} c_n \, r^{-n-1} \, P_n(\cos \theta).$$

The potential must also remain finite at $r = 0$, so the potential inside the sphere has the form

$$V_m^{int} = V_0 + \sum_{n=0}^{\infty} d_n \, r^n P_n(\cos \theta).$$

At the boundary surface $r = a$, $H_t = H_\theta$ must be continuous, so

$$-\sum_{n=0}^{\infty} c_n \, a^{-n-2} \frac{dP_n}{d\theta} = -\sum_{n=0}^{\infty} d_n \, a^{n-1} \frac{dP_n}{d\theta}.$$

Since this must true for any θ, the coefficients must satisfy the relation

$$c_n = d_n a^{2n+1}. \tag{3.72}$$

At the boundary surface $r = a$, $B_n = B_r$ must also be continuous. After simplifying we have

$$\mu_0 \left[H_0 P_1 + \sum_{n=0}^{\infty} c_n(n+1)a^{-n-2} P_n \right] = \mu \left[H_0 P_1 - \sum_{n=0}^{\infty} d_n n a^{n-1} P_n \right].$$

We require that this relation hold for any value of n. For $n = 0$, we obtain

$$\mu_0 \, c_0 \, a^{-2} = 0.$$

This shows that $c_0 = 0$ and from Equation 3.72 we find that $d_0 = 0$. For $n = 1$, we find that

$$\mu_0 (H_0 + 2c_1 a^{-3}) = \mu(H_0 - d_1). \tag{3.73}$$

Substituting Equation 3.72 for c_1, we obtain

$$d_1 = \frac{\mu - \mu_0}{\mu + 2\mu_0} H_0.$$

Substituting this back into Equation 3.72, we find

$$c_1 = \frac{\mu - \mu_0}{\mu + 2\mu_0} a^3 H_0.$$

For $n > 1$, we have

$$\mu_0 \, c_n(n+1)a^{-n-2} = -\mu n d_n \, a^{n-1}. \tag{3.74}$$

Using Equation 3.72 for c_n, we obtain $d_n = 0$. Using this in Equation 3.72, we find $c_n = 0$. Thus the only nonvanishing coefficients are c_1 and d_1. The solution for the potential outside the sphere is

$$V_m^{ext} = -H_0 \, r \cos\theta + \frac{\mu - \mu_0}{\mu + 2\mu_0} \frac{a^3}{r^2} H_0 \cos\theta$$

and the field components are

$$B_r = B_0 \cos \theta \left[1 + 2 \left(\frac{\mu_r - 1}{\mu_r + 2} \right) \frac{a^3}{r^2} \right]$$

$$B_\theta = -B_0 \sin \theta \left[1 - \left(\frac{\mu_r - 1}{\mu_r + 2} \right) \frac{a^3}{r^2} \right].$$

The potential inside the sphere is

$$V_m^{int} = -H_0 \, r \cos \theta + \frac{\mu - \mu_0}{\mu + 2 \mu_0} H_0 \cos \theta$$

and the field components are

$$B_r = \mu H_0 \cos \theta \left[1 - \left(\frac{\mu_r - 1}{\mu_r + 2} \right) \right]$$

$$B_\theta = -\mu H_0 \sin \theta \left[1 - \left(\frac{\mu_r - 1}{\mu_r + 2} \right) \right].$$

The magnetic flux density in the vicinity of a sphere with $\mu_r = 20$ is shown in Figure 3.9. The lines of B are pulled into the sphere and approach the boundary approximately along a normal. The field inside the sphere is parallel to the external field.

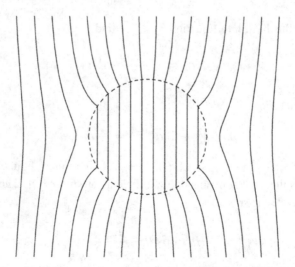

Figure 3.9 Cross-section of the permeable sphere in an external magnetic field.

3.10 Green's theorem

Let V be a region of space enclosed by the surface S. Let ψ and ϕ be scalar functions of position that have continuous first and second derivatives. Applying the divergence theorem to the vector $\psi\nabla\phi$, we get

$$\int \nabla\cdot(\psi\nabla\phi)dV = \int (\psi\nabla\phi)\cdot\hat{n}\ dS. \tag{3.75}$$

Since

$$\nabla\cdot(\psi\nabla\phi) = \nabla\psi\cdot\nabla\phi + \psi\nabla^2\phi$$

and

$$\nabla\phi\cdot\hat{n} = \frac{\partial\phi}{\partial n},$$

Equation 3.75 becomes

$$\int \nabla\psi\cdot\nabla\phi\ dV + \int \psi\nabla^2\phi\ dV = \int \psi\frac{\partial\phi}{\partial n}dS. \tag{3.76}$$

If we repeat this calculation, interchanging the role of ϕ and ψ, we obtain

$$\int \nabla\phi\cdot\nabla\psi\ dV + \int \phi\nabla^2\psi\ dV = \int \phi\frac{\partial\psi}{\partial n}dS. \tag{3.77}$$

Subtracting Equation 3.77 from Equation 3.76, we obtain Green's second identity or *Green's theorem* [19]

$$\int(\psi\nabla^2\phi - \phi\nabla^2\psi)dV = \int \left(\psi\frac{\partial\phi}{\partial n} - \phi\frac{\partial\psi}{\partial n}\right)dS. \tag{3.78}$$

For two-dimensional problems, there is a corresponding Green's theorem on the plane given by [20]

$$\iint \left(\frac{\partial Q}{\partial x} - \frac{\partial P}{\partial y}\right)dx\ dy = \oint (P\ dx + Q\ dy), \tag{3.79}$$

where P and Q are continuous functions of x and y and have continuous partial derivatives.

To apply Green's theorem to magnetostatics, let us choose the function ψ to be proportional to the inverse distance between an element of current at r' and a field observation point at r.

$$\psi = \frac{1}{4\pi R} = \frac{1}{4\pi|\overrightarrow{r} - \overrightarrow{r}'|}. \tag{3.80}$$

In the case where currents are present inside V, we choose ϕ to be one of the components of the vector potential, e.g., A_x. Then we have

$$\nabla^2 A_x = -\mu_0 J_x$$

and from Equation 1.23

$$\nabla^2 \left(\frac{1}{4\pi R}\right) = -\delta(\overrightarrow{r} - \overrightarrow{r}').$$

Substituting this into Green's theorem, we get

$$\int \left[\frac{1}{4\pi R}(-\mu_0 J_x) - A_x\left(-\delta(\overrightarrow{r} - \overrightarrow{r}')\right)\right] dV' = \int \left[\frac{1}{4\pi R}\frac{\partial A_x}{\partial n} - A_x\frac{\partial}{\partial n}\left(\frac{1}{4\pi R}\right)\right] dS'.$$

Because of the delta function, we can solve this equation for $A_x(r)$.

$$A_x(\overrightarrow{r}) = \frac{\mu_0}{4\pi}\int \frac{J_x}{R} dV' + \frac{1}{4\pi}\int \left[\frac{1}{R}\frac{\partial A_x}{\partial n} - A_x\frac{\partial}{\partial n}\left(\frac{1}{R}\right)\right] dS'. \tag{3.81}$$

The volume integral represents a particular solution of the Poisson equation and only includes the effects of currents inside V. Any additional currents outside V influence the value of the surface integral. If we let r be a set of points on the surface S, then this represents an integral equation for the unknown vector potential.

We can use Green's theorem to develop integral equation solutions for Dirichlet and Neumann boundary value problems. We generalize Equation 3.80 used in the previous derivation by defining the *Green's function* [21]

$$G(\overrightarrow{r}, \overrightarrow{r}') = \frac{1}{4\pi|\overrightarrow{r} - \overrightarrow{r}'|} + L(\overrightarrow{r}, \overrightarrow{r}'),$$

where L is an arbitrary solution of the Laplace equation inside V. Substituting this into Green's theorem, we obtain an equation similar to Equation 3.81 with $1/R$ replaced with G. In the case where there are no currents inside V, we have

$$\phi(\overrightarrow{r}) = \frac{1}{4\pi}\int \left[G(\overrightarrow{r}, \overrightarrow{r}')\frac{\partial\phi}{\partial n} - \phi\frac{\partial}{\partial n}G(\overrightarrow{r}, \overrightarrow{r}')\right] dS', \tag{3.82}$$

where ϕ represents either the magnetic scalar potential or one of the components of A.

For Dirichlet boundary value problems, if we choose L such that the Green's function

$$G_D = 0 \tag{3.83}$$

when r' is on the surface, we obtain the integral equation

$$\phi(\vec{r}) = -\frac{1}{4\pi} \int \phi(\vec{r}') \frac{\partial}{\partial n} G_D(\vec{r}, \vec{r}') dS'. \tag{3.84}$$

For Neumann boundary value problems for the exterior of the volume V, if we instead choose L such that

$$\frac{\partial G_N}{\partial n} = 0 \tag{3.85}$$

when r' is on the surface, we obtain the integral equation

$$\phi(\vec{r}) = \frac{1}{4\pi} \int G_N(\vec{r}, \vec{r}') \frac{\partial \phi(\vec{r}')}{\partial n} dS'. \tag{3.86}$$

Thus once we have an appropriate Green's function for a given geometry, the potential at some field point r can be found by integration over the boundary surface. The Green's function is the solution of the Poisson equation for a delta function source.

References

[1] D. Iencinella & G. Matteucci, An introduction to the vector potential, *Eur. J. Phys.* 25:249–256, 2004.
[2] J. D. Jackson, *Classical Electrodynamics*, Wiley, 1962, p. 141.
[3] W. Panofsky & M. Phillips, *Classical Electricity and Magnetism*, 2nd ed., Addison-Wesley, 1962, p. 147–151.
[4] P. Lorrain & D. Corson, *Electromagnetic Fields and Waves*, 2nd ed., Freeman, 1970, p. 305.
[5] J. D. Jackson, *op. cit.*, p. 145–146.
[6] L. Eyges, *The Classical Electromagetic Field*, Dover, 1980, p. 166–167.
[7] P. Lorrain & D. Corson, *op. cit.*, p. 92–94.
[8] L. Eyges, *op. cit.*, p. 138–139.
[9] J. D. Jackson, *op. cit.*, p. 47.
[10] J. B. Marion, *Classical Electromagnetic Radiation*, Academic Press, 1965, p. 58–60.
[11] W. Panofsky & M. Phillips, *op. cit.*, p. 88–89.
[12] W. Smythe, *Static and Dynamic Electricity*, 2nd ed., McGraw-Hill, 1950, p. 170–182.
[13] Ibid., p. 187–198.
[14] W. Panofsky & M. Phillips, *op. cit.*, p. 81–82.
[15] G. Arfken, Mathematical Methods for Physicists, 3rd ed., Academic Press, 1985, p. 680–683.

[16] W. Smythe, *op. cit.*, p. 147–153.
[17] J. D. Jackson, *op. cit.*, p. 15–17.
[18] K. Binns & P. Lawrenson, *Analysis and Computation of Electric and Magnetic Field Problems*, 2nd ed., Pergamon Press, 1973, p. 95–103.
[19] J. Stratton, *Electromagnetic Theory*, McGraw-Hill, 1941, p. 165.
[20] W. Kaplan, *Advanced Calculus*, Addison-Wesley, 1952, p. 239–241.
[21] L. Eyges, *op. cit.*, p. 85–91.

4

Conductor-dominant transverse fields

In this chapter, we consider transverse field configurations that can be approximated as uniform along the z axis. We begin by considering the general form of the solutions to the magnetostatic equations in two dimensions. We next treat the fields produced by line currents, current sheets, and current blocks. We find that multipole errors are introduced when we approximate ideal current distributions with practical conductor configurations. The shapes of the fields discussed here are primarily determined by the location of the conductors. Any iron that may be present only acts to enhance the strength of the field in the magnet aperture. A high-field accelerator magnet is one example of a magnet that produces fields of this type. We conclude with a brief discussion of superconductors and 3D conductor configurations used at the end of these magnets.

4.1 General solution to the Laplace equation in two dimensions

The two-dimensional Laplace's equation in the polar coordinates r and θ is

$$\frac{1}{r} \, \partial_r(r \, \partial_r V) + \frac{1}{r^2} \, \partial_\theta^2 V = 0. \tag{4.1}$$

This is identical with the first two terms in Equation 3.39, so the solutions for the radial and azimuthal dependence of the potential follow from Equations 3.43, 3.44, 3.46, and 3.47. Thus the general solution for Laplace's equation in polar coordinates has the form

$$V(r, \theta) = \sum_{n=1}^{\infty} (C_n r^n + D_n r^{-n}) \, (E_n \cos n\theta + F_n \sin n\theta) + (C_0 \ln r + D_0) \, (E_0 + F_0). \tag{4.2}$$

Example 4.1:   vector potential inside a magnet aperture

The aperture of a magnet is the open area enclosed by the coils. We consider here the general form for the magnetic field in a magnet aperture. We use a polar coordinate system with the origin inside the aperture and let the variable V in Equation 4.2 refer to the z component of the vector potential. In addition, we assume that there are no external fields present outside the magnet and no current filaments inside the aperture. In this case, we can ignore the $n = 0$ terms in Equation 4.2. Since the potential must be finite at $r = 0$, we must have all the coefficients $D_n = 0$. Thus the vector potential inside the aperture has the form

$$A_z(r, \theta) = \sum_{n=1}^{\infty} C_n \, r^n (E_n \cos n\theta + F_n \sin n\theta). \tag{4.3}$$

The magnetic field is given by

$$\vec{B}(r, \theta) = \nabla \times \vec{A} = \hat{r} \frac{1}{r} \partial_\theta A_z - \hat{\theta} \, \partial_r A_z. \tag{4.4}$$

Evaluating the radial field component, we find

$$B_r(r, \theta) = \sum_{n=1}^{\infty} C_n \, r^{n-1} [-E_n \, n \sin n\theta + F_n \, n \cos n\theta]. \tag{4.5}$$

Defining the new coefficients

$$\begin{aligned} A_n &= -n \, C_n \, F_n \\ B_n &= -n \, C_n \, E_n, \end{aligned} \tag{4.6}$$

the radial field inside the magnet aperture can be written as

$$B_r(r, \theta) = \sum_{n=1}^{\infty} r^{n-1} (-A_n \cos n\theta + B_n \sin n\theta). \tag{4.7}$$

The coefficients A_n and B_n describe the multipole field content of the transverse field.[1] Returning to Equation 4.4, the azimuthal field component is

$$B_\theta(r, \theta) = -\sum_{n=1}^{\infty} C_n \, n \, r^{n-1} [E_n \cos n\theta + F_n \sin n\theta].$$

---

[1] *Caveat emptor.* The reader should be aware that a number of different definitions are used in the literature to describe the multipole content of a transverse field.

Using Equation 4.6, the azimuthal field in the magnet aperture is

$$B_\theta(r, \theta) = \sum_{n=1}^{\infty} r^{n-1}(A_n \sin n\theta + B_n \cos n\theta). \tag{4.8}$$

On the midplane of the magnet ($\theta = 0, r = x$), B_x and B_y are given by power series in x

$$\begin{aligned} -B_x(x) &= -B_r(x, 0) = A_1 + A_2 x + A_3 x^2 + \cdots \\ B_y(x) &= B_\theta(x, 0) = B_1 + B_2 x + B_3 x^2 + \cdots \end{aligned} \tag{4.9}$$

Example 4.2: scalar potential inside a magnet aperture
We again begin by considering the general form of the solution of Laplace's equation in polar coordinates given in Equation 4.2. We then specialize to the case for a region containing $r = 0$ and obtain an equation analogous to Equation 4.3.

$$\mu_0 V_m(r, \theta) = \sum_{n=1}^{\infty} G_n r^n (H_n \cos n\theta + I_n \sin n\theta). \tag{4.10}$$

The magnetic field is given by

$$\vec{B}(r, \theta) = -\mu_0 \nabla V_m = -\hat{r}\, \mu_0 \frac{\partial V_m}{\partial r} - \hat{\theta}\, \frac{\mu_0}{r} \frac{\partial V_m}{\partial \theta}.$$

Thus the radial field is

$$B_r = -\sum_{n=1}^{\infty} n\, G_n\, r^{n-1}(H_n \cos n\theta + I_n \sin n\theta). \tag{4.11}$$

The field components calculated from this potential must equal the same quantities calculated from the vector potential. We can make B_r have the same form as Equation 4.5 if we demand that

$$-n\, G_n H_n = n\, C_n F_n = -A_n$$
$$-n\, G_n I_n = -n\, C_n E_n = B_n.$$

Thus we identify

$$\begin{aligned} G_n &= C_n \\ H_n &= -F_n \\ I_n &= E_n. \end{aligned}$$

The scalar potential inside the aperture is then given by[2]

$$\mu_0 V_m(r, \theta) = \sum_{n=1}^{\infty} C_n \, r^n (E_n \sin n\theta - F_n \cos n\theta).$$ (4.12)

The resulting magnetic field components are still given by Equations 4.7 and 4.8.

If the field in some region is known, for example through calculations or measurements, then the multipole field coefficients can be determined using Fourier analysis. Multiplying both sides of Equation 4.8 by cos $m\theta$ and integrating around a circular path, we have

$$\int_0^{2\pi} B_\theta(r, \theta) \, \cos m\theta \, d\theta = \sum_{n=1}^{\infty} r^{n-1} [A_n \, \mathbb{I}_1 + B_n \, \mathbb{I}_2],$$

where for $m \geq 1$ and $n \geq 1$ the integrals have the values[3]

$$\mathbb{I}_1 = \int_0^{2\pi} \sin n\theta \cos m\theta \, d\theta = 0$$ (4.13)

and

$$\mathbb{I}_2 = \int_0^{2\pi} \cos n\theta \cos m\theta \, d\theta = \pi \, \delta_{mn}$$
$$= \int_0^{2\pi} \sin n\theta \sin m\theta \, d\theta.$$ (4.14)

Thus we find one set of multipole field components is given by

$$B_n = \frac{1}{\pi r^{n-1}} \int_0^{2\pi} B_\theta(r, \theta) \cos n\theta \, d\theta.$$ (4.15)

Likewise, we can multiply Equation 4.8 with sin $m\theta$ and find the other set of multipole components is

$$A_n = \frac{1}{\pi r^{n-1}} \int_0^{2\pi} B_\theta(r, \theta) \sin n\theta \, d\theta.$$ (4.16)

---

[2] We will see in the next chapter that this relationship between the vector and scalar potentials follows directly from the Cauchy-Riemann equations.
[3] CRC 497, 502.

A similar analysis using Equation 4.7 for the radial component of the field gives

$$B_n = \frac{1}{\pi r^{n-1}} \int_0^{2\pi} B_r(r, \theta) \sin n\theta \, d\theta \qquad (4.17)$$

and

$$A_n = \frac{-1}{\pi r^{n-1}} \int_0^{2\pi} B_r(r, \theta) \cos n\theta \, d\theta. \qquad (4.18)$$

The strength of the multipole fields provides a measure of the field quality. Limits on the field uniformity are imposed by the application that needs the magnetic field. The presence of harmonics of the desired field limits the size of the useful magnet aperture. Sometimes, when examining the field quality of a magnet, it is more useful to examine the relative magnitude of the multipole coefficients with respect to the coefficient for the desired multipole. Thus for a dipole design, for example, one could calculate the dimensionless quantities

$$b_n = \frac{B_n \, r_0^{n-1}}{B_1},$$

where r_0 is a reference radius, typically ~2/3 of the magnet aperture.

The boundary conditions for the vector potential in polar coordinates at some radius r_b can be determined from the boundary conditions on the magnetic field. From the condition on the normal component of B, we have

$$B_r^{(1)} = B_r^{(2)}$$

$$\frac{1}{r_b} \frac{\partial A_z^{(1)}}{\partial \theta} = \frac{1}{r_b} \frac{\partial A_z^{(2)}}{\partial \theta}.$$

From this relation, we know that $A^{(2)}$ can differ from $A^{(1)}$ by at most a constant, which we can ignore since constants are removed when we take derivatives to obtain the field. Thus we have

$$A_z^{(2)}(r_b, \theta) = A_z^{(1)}(r_b, \theta). \qquad (4.19)$$

From the boundary condition on H_t,

$$H_t^{(2)} - H_t^{(1)} = K,$$

we have

$$-\frac{1}{\mu^{(2)}} \frac{\partial A_z^{(2)}(r_b, \theta)}{\partial r} + \frac{1}{\mu^{(1)}} \frac{\partial A_z^{(1)}(r_b, \theta)}{\partial r} = K. \qquad (4.20)$$

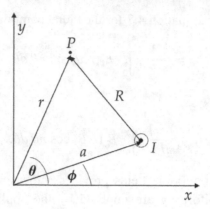

Figure 4.1 Geometry of a line current.

4.2 Harmonic expansion for a line current

Consider a line current perpendicular to the *x-y* plane, as shown in Figure 4.1. The vector potential for a line current was given in Equation 3.8

$$A_z(r, \theta) = -\frac{\mu I}{2\pi} \ln\left(\frac{R}{r_0}\right), \tag{4.21}$$

where the distance from the line current at (a, ϕ) to the observation point P located at (r, θ) is given by

$$R = \left\{ (r \cos \theta - a \cos \phi)^2 + (r \sin \theta - a \sin \phi)^2 \right\}^{1/2} \tag{4.22}$$

and r_0 is some constant reference radius for the two-dimensional potential. We look for a harmonic expansion for A_z. When $r > a$ we extract a factor of r^2 from the logarithm and find that

$$\ln(R) = \tfrac{1}{2} \ln(r^2) + \tfrac{1}{2} \ln\left[1 + \frac{a^2}{r^2} - 2\frac{a}{r} \cos(\theta - \phi) \right].$$

Writing the cosine in terms of complex exponentials gives

$$\ln(R) = \ln(r) + \tfrac{1}{2} \ln\left[1 + \frac{a^2}{r^2} - \frac{a}{r} e^{i(\theta-\phi)} - \frac{a}{r} e^{-i(\theta-\phi)} \right]$$

$$= \ln(r) + \tfrac{1}{2} \left[\ln\left(1 - \frac{a}{r} e^{i(\theta-\phi)} \right) + \ln\left(1 - \frac{a}{r} e^{-i(\theta-\phi)} \right) \right].$$

Using the series expansion for $\ln(1 + x)$ for $x < 1$, we find

$$\ln(R) = \ln(r) - \frac{a}{r} \cos{(\theta - \phi)} - \frac{1}{2} \left(\frac{a}{r}\right)^2 \cos{[2(\theta - \phi)]} + \cdots$$

$$= \ln(r) - \sum_{n=1}^{\infty} \frac{1}{n} \left(\frac{a}{r}\right)^n \cos{[n(\theta - \phi)]}.$$

Thus the vector potential for the line current when $r > a$ is [1]

$$A_z(r, \theta) = -\frac{\mu I}{2\pi} \ln(r) + \frac{\mu I}{2\pi} \sum_{n=1}^{\infty} \frac{1}{n} \left(\frac{a}{r}\right)^n \cos{[n(\theta - \phi)]} \qquad (4.23)$$

plus a constant term involving r_0. The expansion of $\ln R$ for the case $r < a$ can be done in a similar manner by first extracting a factor of a^2 from the argument of the logarithm. This results in the vector potential

$$A_z(r, \theta) = -\frac{\mu I}{2\pi} \ln(a) + \frac{\mu I}{2\pi} \sum_{n=1}^{\infty} \frac{1}{n} \left(\frac{r}{a}\right)^n \cos{[n(\theta - \phi)]}. \qquad (4.24)$$

Example 4.3: line current in circular iron cavity
Consider a line current at radius a inside the circular aperture of a piece of iron with radius R, as shown in Figure 4.2. We know from Equations 4.23 and 4.24 that the contribution to the total vector potential from the line current has different expansions depending on whether r is greater than or lesser than the radius a of the line current. Similarly, the field induced in the iron has different expressions depending on whether r is greater than or smaller than the radius R of the opening in the iron.[2, 3] The induced vector potential must have the general form given in Equation 4.2. To match the field from the line current at the boundaries, the induced field must have the same angular dependence as the line current. The induced field must also be finite at $r = 0$ and at $r \to \infty$. Let $k = I/2\pi$ and $\omega = \theta - \phi$. Then the total vector potential in the region (1) with $r < a$ is

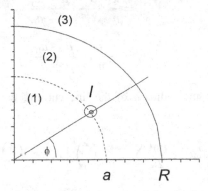

Figure 4.2 Quarter section of a line current in a circular iron cavity.

$$A_{z1} = -\mu_0 k \ln(a) + \mu_0 k \sum_{n=1}^{\infty} \frac{1}{n} \left(\frac{r}{a}\right)^n \cos n\omega + \mu_0 \sum_{n=1}^{\infty} C_n \, r^n \cos n\omega.$$

In the region (2) with $a < r < R$ between the line current and the iron boundary, the vector potential is

$$A_{z2} = -\mu_0 k \ln(r) + \mu_0 k \sum_{n=1}^{\infty} \frac{1}{n} \left(\frac{a}{r}\right)^n \cos n\omega + \mu_0 \sum_{n=1}^{\infty} C_n \, r^n \cos n\omega.$$

Lastly, in the region (3) with $r > R$ inside the iron, the vector potential is

$$A_{z3} = -\mu k \ln(r) + \mu k \sum_{n=1}^{\infty} \frac{1}{n} \left(\frac{a}{r}\right)^n \cos n\omega + \mu \sum_{n=1}^{\infty} D_n \, r^{-n} \cos n\omega,$$

where $\mu = \mu_r \, \mu_0$ is the assumed constant permeability of the iron. We determine the unknown coefficients C_n and D_n by demanding continuity of

$$B_r = \frac{1}{r} \, \partial_\theta A_z$$

$$H_\theta = -\frac{1}{\mu} \, \partial_r A_z$$

at the surface $r = R$ of the iron.

$$\frac{\mu_0 k}{n} \, a^n R^{-n-1} + \mu_0 C_n \, R^{n-1} = \frac{\mu k}{n} \, a^n R^{-n-1} + \mu D_n \, R^{-n-1}$$

$$-C_n \, R^{n-1} = D_n \, R^{-n-1}.$$

This gives two equations in two unknowns, which can be solved to give,

$$C_n = \frac{k}{n} \, \frac{a^n}{R^{2n}} \, \frac{\mu - \mu_0}{\mu + \mu_0}$$

$$D_n = -\frac{k}{n} \, a^n \, \frac{\mu - \mu_0}{\mu + \mu_0}.$$

In the region inside the radius a of the line current, the vector potential can be expressed as

$$A_{z1} = -\frac{\mu_0 I}{2\pi} \ln(a) + \frac{\mu_0 I}{2\pi} \sum_{n=1}^{\infty} \frac{1}{n} \left(\frac{r}{a}\right)^n \left[1 + \alpha \left(\frac{a}{R}\right)^{2n}\right] \cos n\omega, \qquad (4.25)$$

where

$$\alpha = \frac{\mu - \mu_0}{\mu + \mu_0}.$$

The term in brackets shows the enhancement factor due to the iron. For points inside the iron, the solution is

$$A_{z3} = \frac{\mu I}{2\pi} \left[-\ln(r) + \sum_{n=1}^{\infty} \frac{1}{n} \left(\frac{a}{r}\right)^n (1 - \alpha) \cos n\omega \right]. \tag{4.26}$$

We are now in a position to relate the solution of the previous example with the method of images for a circular boundary that we discussed in Section 2.7. For the region interior to the filament, $r < a$, the vector potential can be written in the form

$$A_{z1} = -\frac{\mu_0 I}{2\pi} \ln(a) + \frac{\mu_0 I}{2\pi} \sum_{n=1}^{\infty} \frac{1}{n} \left(\frac{r}{a}\right)^n \cos n\omega + \frac{\mu_0 I}{2\pi} \sum_{n=1}^{\infty} \frac{1}{n} r^n \frac{\alpha}{r_1^n} \cos n\omega, \tag{4.27}$$

where

$$r_1 = \frac{R^2}{a}.$$

Comparing with Equation 4.24, the first two terms give the potential for the true line current at $r = a$. Since in magnetostatics constant terms in the potential have no physical effects, we can arbitrarily add to the potential a constant term

$$-\frac{\mu_0 I}{2\pi} \ln(r_1).$$

Then this term plus the last term in Equation 4.27 give the potential for a line current at r_1 with current αI. Thus the vector potential in region (1) can be written as

$$A_{z1} = A_{LC}(a) + \alpha A_{LC}(r_1),$$

where A_{LC} is the vector potential for a line current. For the region inside the iron, $r > R$, let us define

$$\beta = 1 - \alpha = \frac{2\mu_0}{\mu + \mu_0}.$$

Then Equation 4.26 can be rewritten in the form

$$A_{z3} = \frac{\mu I}{2\pi} \left[-\ln r + \beta \ \ln r - \beta \ \ln r + \beta \sum_{n=1}^{\infty} \frac{1}{n} \left(\frac{a}{r}\right)^n \cos n\omega \right].$$

Using Equation 4.23 for the vector potential of a line current, we find

$$A_{z3} = \frac{\mu I}{2\pi}\left[-(1-\beta)\,\ln r + \beta\left(-\ln r + \sum_{n=1}^{\infty}\frac{1}{n}\left(\frac{a}{r}\right)^{n}\cos n\omega\right)\right]$$

$$= \alpha\,A_{LC}(0) + \beta\,A_{LC}(a).$$

We see that the coefficients α and β are the same as those for the image currents in Equations 2.27–2.30.

4.3 Field for a current sheet

Consider a conductor in the form of an infinitely thin sheet that is uniform in the z direction, as shown in Figure 4.3. Assume here that the current also flows in the z direction. From the Biot-Savart law for a current sheet, Equation 1.13, we have

$$\vec{B} = \frac{\mu_0}{4\pi}\int\frac{\vec{K}\times\vec{R}}{R^3}\,dS$$

Assume the observation point P is in the x-y plane, as shown in Figure 4.4. We have

$$\vec{K} = \frac{dI}{ds}\,\hat{z}$$
$$\rho^2 = r^2 + a^2 - 2\,r\,a\,\cos\,(\theta - \phi)$$
$$R^2 = \rho^2 + z^2$$
$$dS = ds\,dz,$$

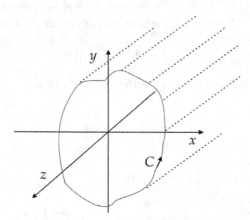

Figure 4.3 Current sheet with cross-section along the curve C and extending infinitely along the z direction.

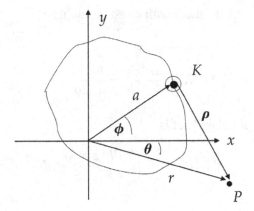

Figure 4.4 Sheet geometry.

where $a = a(s)$, $\phi = \phi(s)$ and s is the arclength around the sheet. The vector ρ is the distance between the current element and the field point in the x-y plane. Thus

$$\vec{B}(r, \theta) = \frac{\mu_0}{4\pi} \int K(s) \int_{-\infty}^{\infty} \frac{\hat{z} \times (\vec{\rho} + z\,\hat{z})}{\{\rho^2 + z^2\}^{3/2}}\, dz\, ds$$

$$= \frac{\mu_0}{4\pi} \int K(s)\ \mathbb{I}(\rho)\ \hat{z} \times \vec{\rho}\ \ ds,$$

where[4]

$$\mathbb{I}(\rho) = \int_{-\infty}^{\infty} \frac{dz}{\{\rho^2 + z^2\}^{3/2}} = \frac{2}{\rho^2}. \tag{4.28}$$

Thus we find that the field from the current sheet is given by

$$\vec{B}(r, \theta) = \frac{\mu_0}{2\pi} \int K(s) \frac{\hat{z} \times \vec{\rho}}{\rho^2}\ ds. \tag{4.29}$$

It will also be useful to have an expression for the vector potential of a current sheet. Assuming the sheet is composed of parallel line currents and using Equation 4.21, we have

$$A_z(r, \theta) = -\frac{\mu_0}{2\pi} \int K(s) \ln\left(\frac{\rho}{\rho_0}\right)\ ds. \tag{4.30}$$

[4] GR 2.271.5.

For the case of a circular sheet with constant radius a,

$$ds = a\, d\phi$$

$$K = \frac{dI}{d\phi}\frac{d\phi}{ds} = \frac{1}{a}\frac{dI}{d\phi}.$$

The field due to the circular sheet is

$$\vec{B}(r,\theta) = \frac{\mu_0}{2\pi}\int_{\phi_1}^{\phi_2} \frac{dI}{d\phi}\, \frac{\hat{z}\times\vec{\rho}}{\rho^2}\, d\phi \qquad (4.31)$$

and the vector potential is

$$A_z(r,\theta) = -\frac{\mu_0}{2\pi}\int_{\phi_1}^{\phi_2}\frac{dI}{d\phi}\, \ln\!\left(\frac{\rho}{\rho_0}\right) d\phi. \qquad (4.32)$$

Example 4.4:   field between two parallel, straight current sheets
Consider the two parallel current sheets shown in Figure 4.5. The current, which is uniform along y, flows into the page on the sheet on the right and returns back out of the page on the sheet on the left. The field observation point P is at (x_o, y_o). We have

$$\vec{\rho} = (x_o - x)\,\hat{x} + (y_o - y)\,\hat{y}$$

$$K = \frac{dI}{dy}.$$

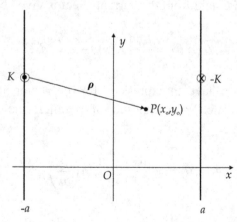

Figure 4.5 Parallel current sheets.

Applying Equation 4.29, the field can be written as

$$\vec{B}(x_o, y_o) = \frac{\mu_0}{2\pi} K \int_{-\infty}^{\infty} \left[-\frac{(x_o - a)\hat{y} - (y_o - y)\hat{x}}{(x_o - a)^2 + (y_o - y)^2} + \frac{(x_o + a)\hat{y} - (y_o - y)\hat{x}}{(x_o + a)^2 + (y_o - y)^2} \right] dy.$$

All the current elements on both sheets give positive B_y because the magnitude of x_o is smaller than a. Thus we can write

$$B_y = \frac{\mu_0}{2\pi} K \left[(a - x_o) \, \mathbb{I}_1 + (a + x_o) \, \mathbb{I}_2 \right],$$

where[5]

$$\mathbb{I}_1 = \int_{-\infty}^{\infty} \frac{dy}{(x_o - a)^2 + (y_o - y)^2} = \frac{\pi}{a - x_o}$$

$$\mathbb{I}_2 = \int_{-\infty}^{\infty} \frac{dy}{(x_o + a)^2 + (y_o - y)^2} = \frac{\pi}{a + x_o}.$$

Thus the vertical field between the sheets is

$$B_y = \mu_0 \frac{dI}{dy}. \tag{4.33}$$

This is, as expected, twice the field we found for a single sheet in Equation 1.17 and independent of the location of the observation point. In a similar manner, we find

$$B_x = \frac{\mu_0}{2\pi} K [y_o \mathbb{I}_1 - \mathbb{I}_3 - y_o \mathbb{I}_2 + \mathbb{I}_4],$$

where

$$\mathbb{I}_3 = \int_{-\infty}^{\infty} \frac{y}{(x_o - a)^2 + (y_o - y)^2} \, dy$$

and $\mathbb{I}_4$ is a similar integral with $x_o + a$ in the denominator. If we let D represent the denominator, then[6]

$$\mathbb{I}_3 = \tfrac{1}{2} \ln[D]_{-\infty}^{\infty} + y_o \mathbb{I}_1.$$

Using l'Hopital's rule, we can show that the first term vanishes, and find that

$$\mathbb{I}_3 = y_o \, \mathbb{I}_1$$
$$\mathbb{I}_4 = y_o \, \mathbb{I}_2.$$

[5] GR 2.172.   [6] GR 2.175.1.

Substituting back in, we find that the horizontal field between the sheets vanishes.

$$B_x = 0.$$

Thus there is a pure dipole field in the region between the current sheets.

4.4 Ideal multipole current sheet

Assume we have a circular current sheet with radius a and with azimuthal current density in the z direction given by

$$\frac{dI}{d\phi} = I_0 \cos m\phi,$$

where I_0 is the current flowing at the midplane ($\phi = 0$). We obtain the vector potential for the sheet by integrating the weighted distribution of the vector potential for a line current. Let us first consider the case where the observation point (r, θ) has $r < a$. Using Equation 4.24 for the line current and ignoring the constant term, the vector potential for the multipole sheet is

$$A_z = \frac{\mu I_0}{2\pi} \sum_{n=1}^{\infty} \frac{1}{n} \left(\frac{r}{a}\right)^n \int_0^{2\pi} \cos m\phi \, \cos[n(\theta - \phi)] \, d\phi.$$

Expanding the integrand and using Equations 4.13 and 4.14, the integral has the value

$$\int_0^{2\pi} \cos m\phi \, \cos[n(\theta - \phi)] \, d\phi = \begin{cases} 0 & \text{if } m \neq n \\ \pi \cos m\theta & \text{if } m = n \end{cases}. \tag{4.34}$$

Therefore the vector potential for $r < a$ is [4]

$$A_z(r, \theta) = \frac{\mu I_0}{2m} \left(\frac{r}{a}\right)^m \cos m\theta \tag{4.35}$$

and the components of the magnetic field are

$$B_r(r, \theta) = -\frac{\mu I_0}{2a} \left(\frac{r}{a}\right)^{m-1} \sin m\theta$$

$$B_r(r, \theta) = -\frac{\mu I_0}{2a} \left(\frac{r}{a}\right)^{m-1} \cos m\theta. \tag{4.36}$$

In the case of a dipole distribution ($m = 1$),

$$B_y = B_r \sin\theta + B_\theta \cos\theta = -\frac{\mu I_0}{2a}$$

$$B_x = B_r \cos\theta - B_\theta \sin\theta = 0$$

showing that the $\cos\theta$ angular distribution also produces a pure vertical field in the magnet aperture.

For the case when $r > a$, we use Equation 4.23 for the vector potential of the line current and obtain

$$A_z = -\frac{\mu I_0}{2\pi} \int_0^{2\pi} \cos m\phi \ \ln r \, d\phi + \frac{\mu I_0}{2\pi} \sum_{n=1}^{\infty} \frac{1}{n} \left(\frac{a}{r}\right)^n \int_0^{2\pi} \cos m\phi \ \cos\left[n(\theta - \phi)\right] d\phi.$$

The first integral vanishes over a complete circle and the second integral can again be evaluated using Equation 4.34. Thus the vector potential of the multipole sheet for the region $r > a$ is

$$A_z(r, \theta) = \frac{\mu I_0}{2m} \left(\frac{a}{r}\right)^m \cos m\theta \qquad (4.37)$$

and the magnetic field is

$$B_r(r, \theta) = -\frac{\mu I_0}{2} \frac{a^m}{r^{m+1}} \sin m\theta$$

$$B_r(r, \theta) = \frac{\mu I_0}{2} \frac{a^m}{r^{m+1}} \cos m\theta. \qquad (4.38)$$

Example 4.5: $\cos m\theta$ sheet in circular iron cavity

Consider a circular $\cos m\theta$ sheet with radius a inside a symmetric circular iron cavity with radius R. We first solve the boundary value problem using vector potentials. There are two vector potentials for the sheet, depending on whether r is smaller than or greater than a, and two vector potentials for the image effects in the iron, depending on whether r is smaller than or greater than R. The angular dependence of the image effects must also use the cosine term in Equation 4.2 and only the mth term in the summation in order to match the boundary conditions. This introduces two unknown coefficients, C and D, and requires that the total vector potential for the three regions be given as

$$A_{z1} = \frac{\mu_0 I_0}{2m} \left(\frac{r}{a}\right)^m \cos m\theta + C \, r^m \cos m\theta$$

$$A_{z2} = \frac{\mu_0 I_0}{2m} \left(\frac{a}{r}\right)^m \cos m\theta + C \, r^m \cos m\theta$$

$$A_{z3} = \frac{\mu I_0}{2m} \left(\frac{a}{r}\right)^m \cos m\theta + D \, r^{-m} \cos m\theta.$$

To determine the values for C and D, we demand that B_r and H_θ are continuous across the iron boundary.

$$\frac{\mu_0 I_0}{2} \left(\frac{a}{R}\right)^m + C\, m\, R^m = \frac{\mu I_0}{2} \left(\frac{a}{R}\right)^m + D\, m\, R^{-m}$$

$$C\frac{R^m}{\mu_0} = -D\frac{R^{-m}}{\mu}.$$

Solving these two equations, we find that the unknown coefficients are

$$C = \frac{\mu_0 I_0}{2\, m\, R^m} \left(\frac{a}{R}\right)^m \frac{\mu - \mu_0}{\mu + \mu_0}$$

$$D = -\frac{\mu I_0}{2\, m\, R^{-m}} \left(\frac{a}{R}\right)^m \frac{\mu - \mu_0}{\mu + \mu_0}.$$

Using these values, the vector potential is now known in the three regions. The effect of the iron can be summarized by defining the iron enhancement factor

$$\alpha_m = 1 + \frac{\mu_r - 1}{\mu_r + 1} \left(\frac{a}{R}\right)^{2m}, \tag{4.39}$$

which agrees with the enhancement factor from Equation 4.25. We can write the vector potential inside the magnet aperture as

$$A_{z1}(r, \theta) = \frac{\mu_0 I_0}{2m} \left(\frac{r}{a}\right)^m \alpha_m \cos m\theta. \tag{4.40}$$

The corresponding field components inside the aperture are

$$B_r(r, \theta) = -\frac{\mu_0 I_0}{2a} \left(\frac{r}{a}\right)^{m-1} \alpha_m \sin m\theta$$

$$B_r(r, \theta) = -\frac{\mu_0 I_0}{2a} \left(\frac{r}{a}\right)^{m-1} \alpha_m \cos m\theta. \tag{4.41}$$

The iron enhancement factor in Equation 4.39 ignores any saturation effects in the iron. When saturation becomes significant, the enhancement factor for the dipole field is decreased. In addition, the saturation of the iron does not occur uniformly. This causes changes in the field from the azimuthal distribution of the enhanced currents, leading to sextupole and higher multipole errors in the field in the magnet aperture. Fortunately, there are techniques, such as modifying the iron shape, which can be used to adjust the value of B_3 at a fixed operating current.[5]

Example 4.6: cos $m\theta$ sheet in circular iron cavity using the scalar potential
It is instructive to use an alternative method of solving the preceding boundary value problem. In this case, we will use the scalar potential and take into account the presence of the current sheet through the addition of another pair of boundary conditions. Unlike the previous example, the unknown coefficients here take into account both the field from the sheet and the field from the images in the iron. We know from the boundary condition across a current sheet, Equation 2.23, that the angular dependence of the current must match the angular dependence of the fields on either side of the sheet. Since the fields are given by the derivative of the potential and the current goes like cos $m\theta$, this implies that we must use the sine term in Equation 4.2 for the potential. Thus the scalar potentials for the three regions are

$$V_{m1} = A\, r^m \sin m\theta$$
$$V_{m2} = (B\, r^m + C\, r^{-m})\, \sin m\theta$$
$$V_{m3} = D\, r^{-m} \sin m\theta.$$

The boundary conditions at the sheet are

$$A\, a^m = B\, a^m - C\, a^{-m}$$
$$-m(B\, a^m + C\, a^{-m}) + m A\, a^m = I_0,$$

while the boundary conditions at the iron surface are

$$-\mu_0(B\, R^m - C\, R^{-m}) = \mu D R^{-m}$$
$$B\, R^m + C\, R^{-m} = D\, R^{-m}.$$

Solving the four equations for the four unknowns, we find

$$A = \frac{I_0}{2ma^m}\, \alpha_m$$

$$B = \frac{I_0\, a^m}{2mR^{2m}}\, \frac{\mu - \mu_0}{\mu + \mu_0}$$

$$C = -\frac{I_0\, a^m}{2m}$$

$$D = -\frac{\mu_0 I_0\, a^m}{m(\mu + \mu_0)}.$$

Evaluating the field components inside the magnet aperture, we again obtain Equation 4.41.

4.5 Multipole dependence on the current distribution

We next seek to determine how the multipole fields in a magnet aperture are related to the current distribution on a circular current sheet.

Recall that the vector potential for a line current at (a, ϕ) is

$$A_z(r, \theta) = \frac{\mu I}{2\pi} \sum_{m=1}^{\infty} \frac{1}{m} \left(\frac{r}{a}\right)^m \cos \left[m(\theta - \phi)\right].$$

The corresponding azimuthal field component is

$$B_\theta(r, \theta) = -\frac{\mu I}{2\pi} \sum_{m=1}^{\infty} \frac{r^{m-1}}{a^m} \cos \left[m(\theta - \phi)\right]. \tag{4.42}$$

From Equation 4.15, the contribution to the normal multipole is

$$B_n = \frac{1}{\pi r^{n-1}} \int_0^{2\pi} B_\theta(r, \theta) \cos n\theta \, d\theta$$

$$= -\frac{\mu I}{2\pi^2} \sum_{m=1}^{\infty} \frac{1}{a^m} \int_0^{2\pi} \cos \left[m(\theta - \phi)\right] \cos n\theta \, d\theta .$$

Expanding the cosine and integrating, we find

$$B_n = -\frac{\mu I}{2\pi^2} \sum_{m=1}^{\infty} \frac{1}{a^m} \pi \cos m\phi \, \delta_{mn}$$

$$= -\frac{\mu I}{2\pi a^n} \cos n\phi . \tag{4.43}$$

For a current sheet, we can generalize this by integrating over the current distribution.

$$B_n = -\frac{\mu}{2\pi a^n} \int_0^{2\pi} \frac{dI}{d\phi} \cos n\phi \, d\phi. \tag{4.44}$$

We can find the skew multipoles in a similar manner using Equation 4.16.

$$A_n = -\frac{\mu}{2\pi a^n} \int_0^{2\pi} \frac{dI}{d\phi} \sin n\phi \, d\phi. \tag{4.45}$$

As the name suggests, a multipole field is characterized by the number of poles it has around the circumference of the sheet. Every magnet has an equal number of positive and negative poles. We use the index N to refer to the design number of pole

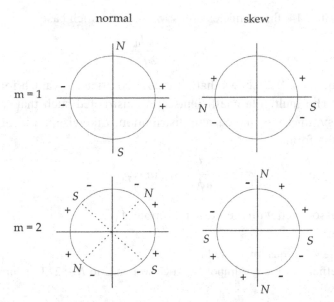

Figure 4.6 Multipole symmetries for dipoles ($m = 1$) and quadrupoles ($m = 2$). N and S indicate the north and south poles of the magnet. Plus and minus signs indicate the direction of the current.

pairs in the multipole magnet. Thus $N = 1$ refers to a dipole magnet because it has one pair of poles, $N = 2$ to a quadrupole, etc. The current in the magnet goes in opposite directions on the adjacent sides of a pole, as shown in Figure 4.6. This corresponds to the way these coils are usually wound, with the cable bending around the pole and returning in the opposite direction. As we can see from Equations 4.44 and 4.45, the multipole coefficients are weighted sums of the current distribution. The B_n coefficients are referred to as the *normal* multipoles. They are largest when the magnitude of the current reaches a maximum on the midplane. The A_n coefficients are referred to as the *skew* multipoles, which are largest when the current changes sign at the midplane. The field in a skew multipole of order N has the same pattern as the normal multipole rotated by $\pi/2N$. For example, a normal dipole has a vertical field, while a skew dipole has a horizontal field.

Consider the current distribution for an ideal multipole of order N given by

$$\frac{dI}{d\phi} = I_0 \cos N\phi.$$

The normal multipole coefficient for a complete angular distribution is

$$B_n = -\frac{\mu_0 I_0}{2\pi \, a^n} \int_{\phi_1}^{\phi_2} \cos N\phi \cos n\phi \, d\phi.$$

From Equation 4.14, this vanishes unless $n = N$, in which case

$$B_N = -\frac{\mu_0 I_0}{2a^N}.$$

Similarly, Equation 4.13 shows that all the A_n coefficients vanish for this current distribution. The multipole coefficients are constructed such that they uniquely identify the symmetry of the current distribution. Likewise, an ideal current distribution of the form

$$\frac{dI}{d\phi} = I_0 \sin N\phi$$

is uniquely associated with the skew multipole A_N.

Example 4.7:   quadrupole field
The coefficient for a quadrupole field corresponds to $N = 2$. The current density is given by

$$\frac{dI}{d\phi} = I_0 \cos 2\phi.$$

Thus

$$B_2 = -\frac{\mu_0 I_0}{2\pi\, a^2} \int_{\phi_1}^{\phi_2} \cos^2(2\phi)\, d\phi$$

$$= -\frac{\mu_0 I_0}{2\, a^2}.$$

4.6 Approximate multipole configurations

The idealized multipole current configurations discussed previously require a continuously varying distribution of current around the entire circumference. On the other hand, actual magnets are typically constructed from multiple layers of cables with uniform current density. Thus it is important to develop methods for approximating the desired multipole distribution, such that it produces the maximum amount of the desired multipole and still meets the required field quality for the magnet. To illustrate this, we consider here the design of a normal dipole magnet using circular current sheet sectors with constant current density. We know that the idealized current distribution for a dipole goes like $\cos\theta$. The simplest approximation to a $\cos\theta$ distribution is to put constant current sectors in each of the four quadrants, as shown in Figure 4.7. The multipole fields result from the sum of the contributions of the four sheets.

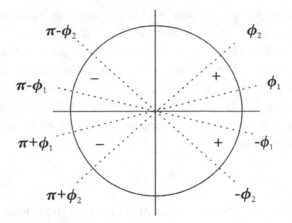

Figure 4.7 Dipole approximation.

$$B_n = -\frac{\mu_0 I}{2\pi a^n} \left\{ \int_{\phi_1}^{\phi_2} \cos n\phi \, d\phi - \int_{\pi-\phi_2}^{\pi-\phi_1} \cdots - \int_{\pi+\phi_1}^{\pi+\phi_2} \cdots + \int_{-\phi_2}^{-\phi_1} \cdots \right\}$$

Let $S(\ ) = \sin(\ )$. Then B_n can be written as the sum of eight sine terms.

$$B_n = -\frac{\mu_0 I}{2\pi n a^n} \left[S(n\phi_2) - S(n\phi_1) - S(n\pi - n\phi_1) + S(n\pi - n\phi_2) \right.$$
$$\left. -S(n\pi + n\phi_2) + S(n\pi + n\phi_1) - S(n\phi_1) + S(n\phi_2) \right].$$

If n is even, the terms in the brackets cancel, so $B_n = 0$. For odd values of n, we get

$$B_n = -\frac{2\mu_0 I}{\pi n a^n} \left[\sin n\phi_2 - \sin n\phi_1 \right]. \tag{4.46}$$

A similar calculation shows that $A_n = 0$ for both odd and even values of n. Thus the allowed harmonics for this current configuration are just the B_n, for odd values of n.

In addition to the desired harmonic B_1 that represents the dipole, the approximation of the ideal multipole distribution above introduces other harmonics, which represent errors to the desired field. The dominant allowed error here is the sextupole term B_3. The angle ϕ_1 is typically set as close to 0 as possible in order to maximize the dipole strength. The angle ϕ_2 can then be used to remove the sextupole component from the field. Since

$$B_3 \simeq -\frac{2\mu_0 I}{3\pi a^3} \sin 3\phi_2,$$

we can eliminate the sextupole term by setting $\phi_2 = \pi/3$. Removing other allowed higher harmonics from the field requires additional degrees of freedom. A number of coil configurations have been proposed to approximate a cos θ distribution.[6] For example, a second current sector can be added in each quadrant that is separated from the first sector by a non-conducting spacer, or sectors could be added at different radii.[7] In principle, these additional sectors could also have independent currents. A two-layer design with spacers is described in Section 11.6.

The allowed harmonics for any multipole of order N follow from the requirement that the direction of the current is in opposite directions on either side of a pole. The first pole is located at

$$\phi = \frac{\pi}{2N}.$$

Referring to Figure 4.7, the current distributions on the opposite sides of the pole are

$$\frac{dI}{d\phi}(\beta) = -\frac{dI}{d\phi}(\phi),$$

where

$$\beta = 2\frac{\pi}{2N} - \phi = \frac{\pi}{N} - \phi.$$

To get a net contribution to a normal multipole of order n

$$B_n \propto \frac{dI}{d\phi}(\phi) \cos n\phi$$

from the current on both sides of the pole, the cosine function must also change sign.

$$\cos\left[n\left(\frac{\pi}{N} - \phi\right)\right] = -\cos n\phi$$

$$\cos\frac{n\pi}{N}\cos n\phi + \sin\frac{n\pi}{N}\sin n\phi = -\cos n\phi.$$

This requires that

$$\cos\frac{n\pi}{N} = -1$$

$$\sin\frac{n\pi}{N} = 0.$$

The sine equation requires that n/N is an integer, while the more restrictive cosine relation demands that n/N is an odd integer. Thus the allowed B_n must have

$$n = N(2m+1), \quad m = 0, 1, 2, \cdots \tag{4.47}$$

Table 4.1 *Multipole symmetries [8]*

Symmetry	Normal multipoles	Skew multipoles	Example
Up-down symmetric		all $A_n = 0$	Normal dipole
Up-down antisymmetric	all $B_n = 0$		Skew dipole
Left-right symmetric	$B_n = 0$ for odd n	$A_n = 0$ for even n	Normal quadrupole
Left-right antisymmetric	$B_n = 0$ for even n	$A_n = 0$ for odd n	Normal dipole

A similar argument shows that the allowed skew multipoles also satisfy Equation 4.47.

The symmetry of the current distribution is directly related to the allowed harmonics. Let $K(\phi) = dI/d\phi$, for example, and assume the current distribution is up-down symmetric, so that

$$K(\phi) = K(-\phi).$$

Then the skew multipoles are

$$A_n = -\frac{\mu_0}{2\pi a^n} \int_0^\pi [K(\phi) \sin n\phi + K(-\phi) \sin (-n\phi)] \, d\phi$$

$$= \frac{\mu_0}{2\pi a^n} \int_0^\pi K(\phi)[\sin n\phi - \sin n\phi] \, d\phi = 0.$$

Thus the fact that the dipole approximation had $A_n = 0$ follows from the up-down symmetry of the current distribution that we used. The consequences of some other symmetries for current distributions are listed in Table 4.1. These symmetries are inevitably violated to some extent in building an actual magnet and this leads to the presence of "nonallowed" multipoles in the fields. Random errors in the construction of the magnet can introduce values for any multipole.[9] If these nondesired multipoles exceed their tolerances, they must be removed by modifications in the manufacturing process or by introducing correction coils.

4.7 Field for a block conductor

Block conductors, which have a finite area in the x-y plane, are the most realistic approximation to actual conductors in two dimensions. We again consider the case where the conductor is infinitely long in the z direction and where the currents only flow along z. From the Biot-Savart law Equation 1.14, we have

$$\vec{B} = \frac{\mu_0}{4\pi} \int \frac{\vec{J} \times \vec{R}}{R^3} \, dV.$$

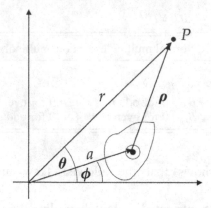

Figure 4.8 Current block geometry.

Assume an observation point P is located at (r, θ) in the x-y plane and the current element is at (a, ϕ), as shown in Figure 4.8. Let the vector ρ be the distance between the current element and the field point in the x-y plane and let σ be the current density in the conductor. Then we have

$$\vec{J} = \sigma \hat{z}$$
$$\vec{R} = \vec{\rho} + z\hat{z}$$
$$dV = dS\ dz.$$

Thus B can be written as

$$\vec{B}(r, \theta) = \frac{\mu_0}{4\pi} \int \sigma \int_{-\infty}^{\infty} \frac{\hat{z} \times [\vec{\rho} + z\hat{z}]}{\{\rho^2 + z^2\}^{3/2}}\ dz\ dS$$

$$= \frac{\mu_0}{4\pi} \int \sigma\, \mathbb{I}(\rho)\, \hat{z} \times \vec{\rho}\ dS,$$

where $\mathbb{I}(\rho)$ is given by Equation 4.28. We find that the field from the current block is given by

$$\vec{B}(r, \theta) = \frac{\mu_0}{2\pi} \int \sigma \frac{\hat{z} \times \vec{\rho}}{\rho^2}\ dS. \qquad (4.48)$$

The vector potential for a current block can be found by integrating the vector potential for the line current, Equation 4.21, over the area of the block.

$$A_z(r, \theta) = -\frac{\mu_0}{2\pi} \int \sigma \ln\left(\frac{\rho}{r_o}\right) dS. \qquad (4.49)$$

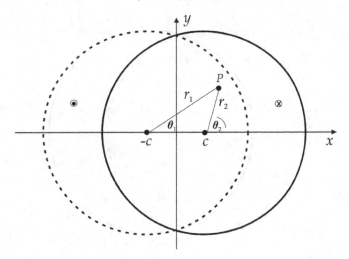

Figure 4.9 Overlapping circular cylindrical conductors.

Example 4.8: overlapping circular conductors
Imagine we have two circular cylindrical conductors with constant current density flowing in opposite directions. We know from Equation 1.27 that the field inside the conductor is

$$B_\phi = \frac{\mu_0 J \rho}{2}.$$

Suppose we overlap the two conductors with the centers displaced along the x axis at c and $-c$, as shown in Figure 4.9. The field at some arbitrary point P in the overlap region is the sum of the fields from the two conductors. From the geometry in the figure, we see that [10]

$$B_x = \frac{\mu_0 J}{2}[-r_2 \sin \theta_2 + r_1 \sin \theta_1] = 0$$

$$B_y = \frac{\mu_0 J}{2}[r_1 \cos \theta_1 - r_2 \cos \theta_2] = \frac{\mu_0 J}{2} 2c,$$

where $2c$ is the separation between the centers of the two circles. Thus the field in the overlap region is a pure dipole. The strength of the field is proportional to the separation between the circles. In the region where the two coils overlap, the net current is zero. Thus the conductor in the overlap region can be removed without affecting the field there.

Next we examine the coil thickness t as a function of the azimuthal angle ϕ, as shown in Figure 4.10. The coil thickness at some angle ϕ is the distance between the points P_1 and P_2. Point P_1 is determined by the intersection of circle 1

$$(x + c)^2 + y^2 = a^2$$

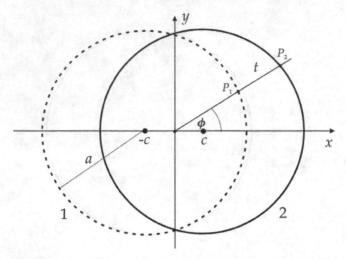

Figure 4.10 Conductor thickness for overlapping circle configuration.

with the straight line

$$y = x \tan \phi = mx.$$

We find that

$$x_1 = \frac{-c + \sqrt{a^2 - m^2 c^2 + m^2 a^2}}{1 + m^2}.$$

Point P_2 is determined by the intersection of circle 2

$$(x - c)^2 + y^2 = a^2$$

with the straight line. We find that the expression for x_2 is the same as the one for x_1, except that the first term in the numerator is $+c$ instead of $-c$. Thus we have

$$\Delta x = x_2 - x_1 = \frac{2c}{1 + \tan^2 \phi}$$

$$\Delta y = y_2 - y_1 = \Delta x \, \tan \phi.$$

The resulting thickness of the conductor is

$$t(\phi) = \sqrt{(\Delta x)^2 + (\Delta y)^2}$$

$$= 2c \cos \phi.$$

Thus the overlapping circular conductors represent another form of a cosine current distribution. Quadrupole fields can be designed in the same manner using overlapping elliptical conductors.[8]

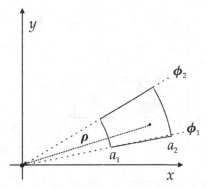

Figure 4.11 Annular current block.

Example 4.9: on-axis field of an annular sector with constant current density
Consider the annular sector conductor shown in Figure 4.11. For a field point on the axis we have

$$\overrightarrow{\rho} = -a \cos \phi \, \hat{x} - a \sin \phi \, \hat{y}.$$

Then Equation 4.48 gives

$$\overrightarrow{B}(0,0) = -\frac{\mu_0 \sigma}{2\pi} \int_{a_1}^{a_2} \int_{\phi_1}^{\phi_2} \frac{\hat{z} \times (a \cos \phi \, \hat{x} + a \sin \phi \, \hat{y})}{a^2} \, a \, d\phi \, da$$

$$= -\frac{\mu_0 \sigma}{2\pi} \int_{a_1}^{a_2} \int_{\phi_1}^{\phi_2} (\cos \phi \, \hat{y} - \sin \phi \, \hat{x}) \, d\phi \, da.$$

Thus the on-axis field of the annular sector is

$$\overrightarrow{B}(0,0) = -\frac{\mu_0 \sigma}{2\pi} (a_2 - a_1) [(\sin \phi_2 - \sin \phi_1)\hat{y} + (\cos \phi_2 - \cos \phi_1)\hat{x}].$$

$$(4.50)$$

We see that the field for the case of constant current density is directly proportional to the radial thickness of the conductor.

Example 4.10: field due to a rectangular conductor
Consider a rectangular conductor with constant current density σ in the z direction, as shown in Figure 4.12. We substitute

$$\overrightarrow{\rho} = (x_o - x) \, \hat{x} + (y_o - y) \, \hat{y}$$

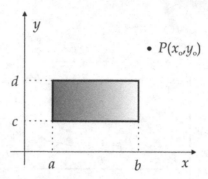

Figure 4.12 Rectangular conductor.

into Equation 4.48 and find that

$$\overrightarrow{dB} = \frac{\mu_0 \sigma}{2\pi} \frac{[(x_o - x)\,\hat{y} - (y_o - y)\,\hat{x}]}{(x_o - x)^2 + (y_o - y)^2}\,dS. \tag{4.51}$$

Looking at B_y, we have

$$B_y = \frac{\mu_0 \sigma}{2\pi} \int_a^b \int_c^d \frac{(x_o - x)}{(x_o - x)^2 + (y_o - y)^2}\,dx\,dy.$$

Integrating first over x, we evaluate[7]

$$\mathbb{I}_1(x_1, x_2) = \int_{x_1}^{x_2} \frac{(x_o - x)}{(x_o - x)^2 + (y_o - y)^2}\,dx$$

$$= -\frac{1}{2}\,\ln[(x_o - x)^2 + (y_o - y)^2]_{x_1}^{x_2}.$$

Then integrating over y, we find[8]

$$\mathbb{I}_2(y_1, y_2, \alpha) = \int_{y_1}^{y_2} \ln[(x_o - \alpha)^2 + (y_o - y)^2]\,dy$$

$$= \left\{ -(y_o - y)\ln[(x_o - \alpha)^2 + (y_o - y)^2] + 2(y_o - y) - 2s(x_o - \alpha)\tan^{-1}\left[\frac{y_o - y}{s(x_o - \alpha)}\right] \right\}_{y_1}^{y_2},$$

where $s = \pm 1$. In order to get a physical solution, we need to choose $s = +1$ when $x_o > \alpha$ and $s = -1$ when $x_o < \alpha$. This is equivalent to taking the absolute value of $(x_o - \alpha)$. Thus we can write B_y as

---

[7] GR 2.175.1.   [8] GR 2.733.1.

$$B_y = -\frac{\mu_0 \sigma}{4\pi} \left[\mathbb{I}_2(c,d,b) - \mathbb{I}_2(c,d,a) \right]$$

$$= -\frac{\mu_0 \sigma}{4\pi} \left\{ (y_o - c) \ \ln[(x_o - b)^2 + (y_o - c)^2] + 2|x_o - b| \ \tan^{-1}\left[\frac{y_o - c}{|x_o - b|}\right] \right.$$

$$- (y_o - d) \ \ln[(x_o - b)^2 + (y_o - d)^2] - 2|x_o - b| \ \tan^{-1}\left[\frac{y_o - d}{|x_o - b|}\right]$$

$$- (y_o - c) \ \ln[(x_o - a)^2 + (y_o - c)^2] - 2|x_o - a| \ \tan^{-1}\left[\frac{y_o - c}{|x_o - a|}\right]$$

$$\left. + (y_o - d) \ \ln[(x_o - a)^2 + (y_o - d)^2] + 2|x_o - a| \ \tan^{-1}\left[\frac{y_o - d}{|x_o - a|}\right] \right\}.$$

From the symmetry of Equation 4.51, we see that the result for B_x is the negative of the result for B_y with the substitutions

$$x \leftrightarrow y$$
$$x_o \leftrightarrow y_o$$
$$(a, b) \leftrightarrow (c, d).$$

Thus B_x is given by

$$B_x = \frac{\mu_0 \sigma}{4\pi} \left\{ (x_o - a) \ \ln[(x_o - a)^2 + (y_o - d)^2] + 2|y_o - d| \ \tan^{-1}\left[\frac{x_o - a}{|y_o - d|}\right] \right.$$

$$- (x_o - b) \ \ln[(x_o - b)^2 + (y_o - d)^2] - 2|y_o - d| \ \tan^{-1}\left[\frac{x_o - b}{|y_o - d|}\right]$$

$$- (x_o - a) \ \ln[(x_o - a)^2 + (y_o - c)^2] - 2|y_o - c| \ \tan^{-1}\left[\frac{x_o - a}{|y_o - c|}\right]$$

$$\left. + (x_o - b) \ \ln[(x_o - b)^2 + (y_o - c)^2] + 2|y_o - c| \ \tan^{-1}\left[\frac{x_o - b}{|y_o - c|}\right] \right\}.$$

Using these expressions for the field, we show in Figure 4.13 a scan of the vertical field component along the x axis for a square conductor centered at the origin. The calculation using these equations fails when the observation point is located at one of the four corners of the rectangle.

We can find the multipoles produced by an annular sector conductor block analogously to the procedure used for current sheets in Section 4.5. The normal multipoles are given by

$$B_n = -\frac{\mu}{2\pi} \iint J(a, \phi) \frac{\cos n\phi}{a^{n-1}} \ da \, d\phi. \tag{4.52}$$

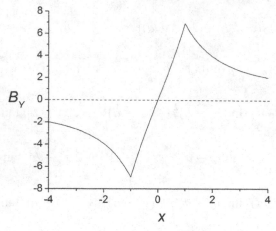

Figure 4.13 Vertical field from a square conductor with sides of length 2 centered at the origin.

The skew multipoles can be found in a similar manner and are given by

$$A_n = \frac{\mu}{2\pi} \iint J(a, \phi) \, \frac{\sin n\phi}{a^{n-1}} \, da \, d\phi. \tag{4.53}$$

4.8 Ideal multipole current block

Consider an annular current block with a pure multipole current density J. We assume that the block is composed of a radial distribution of ideal multipole current sheets extending from a_1 to a_2. We find the vector potential for the block by integrating the potential for a multipole sheet. The current density in the block is assumed to be proportional to $\cos m\phi$.

There are three cases, depending on the radius of the field observation point r.

Case 1: $r < a_1$

The vector potential, Equation 4.35, for the ideal multipole sheet is

$$A_{zsh}(r, \theta) = \frac{\mu I_0}{2m} \left(\frac{r}{a}\right)^m \cos m\theta.$$

Then the vector potential for the current block is

$$A_z(r, \theta) = \frac{\mu J_0}{2m} \cos m\theta \int_{a_1}^{a_2} \left(\frac{r}{a}\right)^m a \, da.$$

For $m \neq 2$, this can be written as

$$A_z(r, \theta) = \frac{\mu J_0}{2m} \, r^m \cos m\theta \left(\frac{a_2^{-m+2} - a_1^{-m+2}}{-m+2} \right). \tag{4.54}$$

When $m = 2$, we have

$$A_z = \frac{\mu J_0 r^2}{4} \cos 2\theta \ln\left(\frac{a_2}{a_1} \right). \tag{4.55}$$

Case 2: $r > a_2$

When the observation point is always larger than the radius of any of the multipole sheets, we use Equation 4.37 for the vector potential of the ideal multipole sheets.

$$A_{zsh}(r, \theta) = \frac{\mu I_0}{2m} \left(\frac{a}{r} \right)^m \cos m\theta.$$

The vector potential of the current block is

$$A_z(r, \theta) = \frac{\mu J_0}{2m \, r^m} \cos m\theta \int_{a_1}^{a_2} a^m \, a \, da.$$

Evaluating the integrals, we have

$$A_z(r, \theta) = \frac{\mu J_0}{2m} \, r^{-m} \cos m\theta \left(\frac{a_2^{m+2} - a_1^{m+2}}{m+2} \right). \tag{4.56}$$

Case 3: $a_1 < r < a_2$

In the third case, where the observation point can be inside the current block, we must break the radial integration into two parts, depending on the relative positions of r and the multipole sheet.

$$A_z(r, \theta) = \frac{\mu J_0}{2m} \cos m\theta \left[r^{-m} \int_{a_1}^{r} a^{m+1} \, da + r^m \int_{r}^{a_2} a^{-m+1} \, da \right].$$

Evaluating the integrals, we have for $m \neq 2$,

$$A_z = \frac{\mu J_0}{2m} \cos m\theta \left[r^{-m} \left(\frac{r^{m+2} - a_1^{m+2}}{m+2} \right) + r^m \left(\frac{a_2^{-m+2} - r^{-m+2}}{-m+2} \right) \right]. \tag{4.57}$$

For the case $m = 2$, we have instead

$$A_z = \frac{\mu J_0}{4} \cos 2\theta \left[r^{-2} \left(\frac{r^4 - a_1^4}{4} \right) + r^2 \ln\left(\frac{a_2}{r} \right) \right]. \tag{4.58}$$

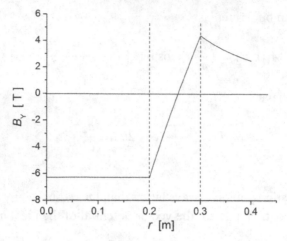

Figure 4.14 Radial scan of the vertical component of the magnetic field of an ideal dipole ($m = 1$, $a_1 = 0.2$ m, $a_2 = 0.3$ m, $J_0 = 100$ A/mm^2).

The components of the magnetic field in each of these regions is easily computed from these vector potentials. Figure 4.14 shows a radial scan of the vertical component of the magnetic field along the midplane for an ideal dipole. The current block extends from 0.2 m to 0.3 m in this example. The field is constant and purely vertical inside the aperture, as expected. The field changes direction inside the current block. The strength of the field slowly falls off outside the block.

4.9 Field from a magnetized body

Let us consider the field intensity outside a magnetized body.[9] We saw from Equation 3.32 that the scalar potential for a magnetic body is

$$V_m = \frac{1}{4\pi} \int \frac{\overrightarrow{M'} \cdot \overrightarrow{R}}{R^3} \, dV'.$$

In the two-dimensional case, let us assume M only has x and y components. The scalar potential is

$$V_m = \frac{1}{4\pi} \int \left[M_x' \int_{-\infty}^{\infty} \frac{R_x}{R^3} \, dz' + M_y' \int_{-\infty}^{\infty} \frac{R_y}{R^3} \, dz' \right] dS',$$

---

[9] We will consider H inside a magnetized body in Chapter 9.

where $R_x = x - x'$, etc. Define the transverse distance in the plane with $z = 0$ as

$$\vec{\rho} = R_x \,\hat{x} + R_y \,\hat{y}.$$

The integrals over z' are[10]

$$\int_{-\infty}^{\infty} \frac{R_x}{R^3}\, dz' = R_x \int_{-\infty}^{\infty} \frac{dz'}{\{\rho^2 + z'^2\}^{3/2}} = \frac{2R_x}{\rho^2}.$$

Then the two-dimensional potential is given by

$$V_m = \frac{1}{2\pi} \int \frac{\vec{M'} \cdot \vec{\rho}}{\rho^2}\, dS'. \tag{4.59}$$

The two-dimensional field is

$$\vec{H}_m = -\frac{1}{2\pi}\, \nabla \int \frac{\vec{M'} \cdot \vec{\rho}}{\rho^2}\, dS'.$$

Using the vector relation B.2,

$$\nabla\left(\vec{M'} \cdot \frac{\vec{\rho}}{\rho^2}\right) = \vec{M'} \times \left(\nabla \times \frac{\vec{\rho}}{\rho^2}\right) + \frac{\vec{\rho}}{\rho^2} \times \left(\nabla \times \vec{M'}\right) + \left(\vec{M'} \cdot \nabla\right)\frac{\vec{\rho}}{\rho^2}$$
$$+ \left(\frac{\vec{\rho}}{\rho^2} \cdot \nabla\right)\vec{M'}. \tag{4.60}$$

The gradient operator only acts on unprimed coordinates, so the second and fourth terms on the right-hand side vanish.

In the first term, we have using Equation B.6

$$\nabla \times \frac{\vec{\rho}}{\rho^2} = \frac{1}{\rho^2}\, \nabla \times \vec{\rho} + \nabla\left(\frac{1}{\rho^2}\right) \times \vec{\rho}. \tag{4.61}$$

The first term here vanishes because ρ is radial and so its curl vanishes. In the second term

$$\nabla\left(\frac{1}{\rho^2}\right) = \hat{x}\, \partial_x\left(\frac{1}{\rho^2}\right) + \hat{y}\, \partial_y\left(\frac{1}{\rho^2}\right)$$
$$= -\frac{2}{\rho^4}\, \vec{\rho}.$$

Thus the cross-product of the last two factors in Equation 4.61 is 0, and the second term also vanishes.

---

[10] GR 2.271.5.

Thus only the third term in Equation 4.60 survives and we have

$$\vec{H}_m = -\frac{1}{2\pi} \int (\vec{M'} \cdot \nabla) \frac{\vec{\rho}}{\rho^2} \, dS'.$$ (4.62)

Expanding the dot product,

$$\vec{H}_m = -\frac{1}{2\pi} \int \left[M_x' \, \partial_x \left(\frac{\vec{\rho}}{\rho^2} \right) + M_y' \, \partial_y \left(\frac{\vec{\rho}}{\rho^2} \right) \right] dS'.$$

Writing out the vector ρ in terms of its x and y components, taking the derivatives, combining terms, and dropping the primes, we find that the two-dimensional field from an iron element is given by

$$\vec{H}_m = -\frac{1}{2\pi} \int \left[\frac{\vec{M}}{\rho^2} - \frac{2(\vec{M} \cdot \vec{\rho})}{\rho^4} \, \vec{\rho} \right] dS.$$ (4.63)

4.10 Superconductors

High-field magnets are usually energized using superconducting cables. In the superconducting state, the resistivity vanishes, so a large current can flow through the magnet coils without losing power due to Joule heating. In a superconducting material, attractive forces between pairs of electrons are transmitted through vibrations in the material lattice.[11] The operating conditions for superconductivity lie below the surface of a three-dimensional space of temperature, magnetic field, and current density. The limiting values on each of the three axes are called the *critical* values. When the superconductor is not in the superconducting state, it is said to be in the *normal* state. There are several classes of superconducting materials. Type I materials exhibit the *Meissner effect*, where any external magnetic field is excluded from the interior of the superconductor. A magnetization is generated in the superconductor that just cancels the external field. This remains true as the external field is increased until it reaches the critical field H_c. Type I superconductors are perfect diamagnetic materials.

Some alloys of intermetallic compounds form what are called type II super-conductors. Two important examples are NbTi and Nb_3Sn. The critical current density for these materials at 4 K is shown as a function of the magnetic flux density in Figure 4.15.[12] NbTi is a useful material for fields up to ~9 T at 4 K, whereas Nb_3Sn can be used up to ~22 T. These materials have two critical magnetic fields, H_{c1} and H_{c2}, which can be much larger than H_{c1}. For $H < H_{c1}$, the material

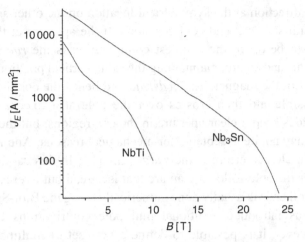

Figure 4.15 Engineering current density at 4 K as a function of the magnetic field.

completely excludes the external flux and it behaves like a type I superconductor. For $H_{c1} < H < H_{c2}$, flux begins to penetrate into the material and the magnitude of the magnetization begins to fall, but electrically it still has zero resistivity. The magnetic flux enters the superconductor in the form of discrete, quantized flux lines known as *fluxoids*. The field still vanishes in the material surrounding the fluxoids. The fluxoids can move because of Lorentz forces, creating heat. To stop the fluxoids from moving, inhomogeneities known as *pinning centers* must be introduced into the lattice. Finally when $H > H_{c2}$, the material returns to the normal state.

Magnets with superconducting cables must deal with the problem of *persistent currents*.[13] As the current in the magnet is ramped up or down, eddy currents[11] are induced in the superconductor. These induced shielding currents produce a field that opposes the change in the field caused by the magnet's power supply. Because of the lack of resistance in a superconductor, the decay times for the induced currents are very long. The persistent currents produce undesired sextupole and higher multipoles that can be particularly significant at low field values.

4.11 End fields

Although this chapter has mainly been concerned with transverse fields that are uniform along the z direction, real currents exist as closed loops, so we must comment on what happens at the end of this type of magnet. At the ends of a dipole, the conductor on one side of a pole must bend in such a way that it returns

---

[11] Eddy currents will be discussed in more detail in Section 10.4.

in the opposite direction at the symmetrical location on the other side of the pole. There are several standard end configurations. If the aperture in the end regions does not need to be open, the simplest configuration is the *racetrack* coil.[10] The coil bends around an arc, maintaining the same vertical position as the coils in the straight part of the magnet. The *bedstead* end bends the conductors by 90° as quickly as possible and then crosses over the pole at a fixed z location. This configuration does keep a clear aperture in the end regions, but the sharp bend in the conductor may not be acceptable for mechanical reasons. Another type of end that maintains a clear aperture is the *saddle* end.[10] In this case, the conductor turns cross over the pole following an arc that is spread out over z. The magnetic design of real coil ends is usually done numerically using the Biot-Savart equation.

The end turns introduce additional multipole contributions to the field of a magnet. However, it is possible to define a new set of multipole coefficients defined in terms of the field components integrated along the axis of the magnet. Flux theorems have been developed that can relate these integrated multipoles to the geometry of the end turns.[14] The z locations where the cross-over for various conductors begin can be adjusted using spacers to help balance the integrated multipoles in the magnet.[15]

References

[1] W. Smythe, *Static and Dynamic Electricity*, 2nd ed., McGraw-Hill, 1950, p. 65.
[2] R. Gupta, *Field Calculations and Computations*, School at Centre for Advanced Technology, Indore, India, 1998, p. 32.
[3] N. Schwerg & C. Vollinger, Analytic models for the calculation of the iron yoke contribution in superconducting accelerator magnets, CERN Accelerator Technology Department report CERN/AT 2007-33, 2007.
[4] P. Schmuser, Superconducting magnets for particle accelerators, in M. Month & M. Dienes (eds.), *The Physics of Particle Accelerators*, AIP Conf. Proc. 249, vol. 2, 1992, p. 1106.
[5] G. Morgan, Shaping of magnetic fields in beam transport magnets, in M. Month & M. Dienes (eds.), *The Physics of Particle Accelerators*, AIP Conf. Proc. 249, vol. 2, 1992, p. 1256–1259.
[6] M. Wilson, *Superconducting Magnets*, Oxford University Press, 1983, p. 32.
[7] J. Coupland, Dipole, quadrupole and higher order fields from simple coils, *Nuc. Instr. Meth.* 78:181, 1970.
[8] A. Jain, Basic theory of magnets, in S. Turner (ed.), *CERN Accelerator School on Measurement and Alignment of Accelerator and Detector Magnets*, CERN 98–05, 1998, p. 25.
[9] G. Parzen, Random errors in the magnetic field of superconducting dipoles and quadrupoles, *Part. Acc.* 6:237, 1975.
[10] M. Wilson, *op. cit.*, p. 27–28.
[11] Ibid., p. 279.
[12] https://nationalmaglab.org/magnet-development/applied-superconductivity-center.

[13] P. Schmuser, *op. cit.*, p. 1123–1141.

[14] F. Mills & G. Morgan, A flux theorem for the design of magnet coil ends, *Part. Acc.* 5:227, 1973.

[15] R. Palmer & A. Tollestrup, Superconducting magnet technology for accelerators, in J. Jackson, H. Gove & R. Schwitters (eds.), *Annual Review of Nuclear and Particle Physics*, vol. 34, Annual Reviews Inc., 1984, p. 269–270.

5

Complex analysis of transverse fields

In this chapter, we continue the discussion of transverse fields that are determined by the location of the conductors. In the "central" region, far from the magnet ends, the powerful methods of complex analysis[1] can be applied to the calculation of potentials, magnetic fields, multipoles and forces. Many of the topics in this chapter are based on a series of important papers by Richard Beth and by Klaus Halbach. Beginning with the field from a line current, we consider methods for calculating the fields from current sheets. Then we use the complex form of Green's theorem to express the fields of block conductors in terms of contour integrals.

5.1 Complex representation of potentials and fields

We define the complex potential function as

$$W(z) = u(x,y) + iv(x,y).$$

The real and imaginary parts of W must satisfy the *Cauchy-Riemann equations*, which are expressed in Cartesian coordinates as

$$\frac{\partial u}{\partial x} = \frac{\partial v}{\partial y}$$
$$\frac{\partial u}{\partial y} = -\frac{\partial v}{\partial x}. \tag{5.1}$$

From Equations 3.2 and 3.26 in free space in two dimensions, we have

$$B_x = \frac{\partial A_z}{\partial y} = -\mu_0 \frac{\partial V_m}{\partial x}$$
$$B_y = -\frac{\partial A_z}{\partial x} = -\mu_0 \frac{\partial V_m}{\partial y}. \tag{5.2}$$

---

[1] A brief summary of some important results from the theory of complex variables is given in Appendix E.

These equations can be put into the form of the Cauchy-Riemann equations by associating

$$u = A_z$$
$$v = \mu_0 V_m.$$

Thus in two dimensions, the vector and scalar potentials are related to each other as the real and imaginary parts of the complex potential function

$$W(z) = A_z + i\mu_0 V_m. \tag{5.3}$$

Now consider the derivative of the complex potential. A complex function with a continuous derivative is known as an *analytic function*. A complex derivative must give the same result independent of the manner that Δz approaches 0. In the case when $\Delta z = \Delta x$, we have

$$\frac{dW}{dz} = \frac{\partial W}{\partial x} = \frac{\partial A_z}{\partial x} + i\mu_0 \frac{\partial V_m}{\partial x} = -B_y - iB_x.$$

If we had chosen $\Delta z = i\Delta y$ instead, we would obtain the same expression for B. So in either case we find

$$i\frac{dW}{dz} = B_x - iB_y.$$

Defining the complex magnetic field as[2]

$$B(z) = B_x + iB_y, \tag{5.4}$$

we find the relation between the magnetic field and the potential is

$$B^*(z) = i\frac{dW}{dz}, \tag{5.5}$$

where B^* is the complex conjugate of B.[1]

We can transform the magnetic field between Cartesian and polar coordinates by using the complex rotation variable. Let

$$B_c = B_x + iB_y$$
$$B_p = B_r + iB_\theta$$

---

[2] Unfortunately, Beth and Halbach use different definitions for the complex magnetic field H and use different systems of units, so some care must be exercised in comparing their results. We follow Halbach's conventions here in defining the components of H the same way as normal complex variables and using the SI system of units.

be the Cartesian and polar representations of a complex variable. Defining

$$R = e^{i\theta} = \cos\theta + i\sin\theta,$$

we can transform between the two representations using

$$\begin{aligned} B_p &= R^* B_c \\ B_c &= R B_p. \end{aligned} \qquad (5.6)$$

Consider the analytic function

$$f(z) = u(x,y) + iv(x,y). \qquad (5.7)$$

Differentiating the first Cauchy-Riemann Equation 5.1 with respect to x, we have

$$\frac{\partial^2 u}{\partial x^2} = \frac{\partial^2 v}{\partial x \partial y}.$$

The fact that the second partial derivative has to exist follows from the analytic nature of $f(z)$.[2] Differentiating the second Cauchy-Riemann equation with respect to y gives

$$\frac{\partial^2 u}{\partial y^2} = -\frac{\partial^2 v}{\partial x \partial y}.$$

Combining these equations, we find that

$$\frac{\partial^2 u}{\partial x^2} + \frac{\partial^2 u}{\partial y^2} = 0.$$

Thus $u(x,y)$ satisfies the Laplace equation. Similarly, we can differentiate the first Cauchy-Riemann equation with y and the second with x to show that $v(x,y)$ also satisfies the Laplace equation. It follows that the real and imaginary parts of any analytic function satisfy the Laplace equation.

Returning to Equation 5.7, consider the two curves

$$\begin{aligned} u(x,y) &= \alpha_1 \\ v(x,y) &= \beta_1, \end{aligned}$$

where α_1 and β_1 are fixed values. Differentiating u with respect to x, we find

$$\frac{\partial u}{\partial x} + \frac{\partial u}{\partial y}\frac{dy}{dx} = 0.$$

The slope of the curve is

$$m_\alpha = \frac{dy}{dx} = -\frac{\partial_x u}{\partial_y u}.$$

Differentiating v with respect to x, we find

$$\frac{\partial v}{\partial x} + \frac{\partial v}{\partial y}\frac{dy}{dx} = 0$$

and the slope of this curve is

$$m_\beta = -\frac{\partial_x v}{\partial_y v}.$$

The product of the slopes is

$$m_\alpha m_\beta = \frac{\partial_x u}{\partial_y u}\frac{\partial_x v}{\partial_y v}.$$

Rewriting the numerator using the Cauchy-Riemann equations, we find

$$m_\alpha m_\beta = \frac{(\partial_y v)(-\partial_y u)}{\partial_y u \partial_y v} = -1.$$

From analytic geometry, this is the condition that indicates that two lines are perpendicular. Thus the real and imaginary parts of an analytic function describe orthogonal curves. This indicates in particular that the equipotential lines for A_z and V_m cross at right angles.

5.2 Maxwell's equations in complex conjugate coordinates

Instead of defining complex variables as functions of x and y, it is sometimes more convenient to use z and z^* as the independent variables. These are known as *complex conjugate coordinates*.[3] We can write the partial derivatives with respect to x and y in terms of these variables as

$$\frac{\partial}{\partial x} = \frac{\partial}{\partial z} + \frac{\partial}{\partial z^*}$$

$$\frac{\partial}{\partial y} = i\left(\frac{\partial}{\partial z} - \frac{\partial}{\partial z^*}\right). \tag{5.8}$$

The corresponding derivatives with respect to z and z^* are

$$2\frac{\partial}{\partial z} = \frac{\partial}{\partial x} - i\frac{\partial}{\partial y}$$

$$2\frac{\partial}{\partial z^*} = \frac{\partial}{\partial x} + i\frac{\partial}{\partial y}. \tag{5.9}$$

We can use Equation 5.8 to write the magnetostatic Maxwell equations in complex coordinates. The divergence equation

$$\partial_x H_x + \partial_y H_y = 0$$

becomes [1]

$$\frac{\partial H}{\partial z} + \frac{\partial H^*}{\partial z^*} = 0, \tag{5.10}$$

where $H = H_x + iH_y$ is the complex magnetic field intensity. The curl equation

$$\partial_x H_y - \partial_y H_x = J_z \equiv \sigma$$

can be transformed using Equation 5.8 into the form

$$-i\partial_z H + i\partial_{z*} H^* = \sigma.$$

This can be further simplified using Equation 5.10, resulting in two forms for the curl equation.[1]

$$2i\frac{\partial H^*}{\partial z^*} = \sigma$$

$$-2i\frac{\partial H}{\partial z} = \sigma. \tag{5.11}$$

Operating on the complex potential in Equation 5.3, we find

$$2\frac{\partial W}{\partial z^*} = (\partial_x A_z - \partial_y \mu_0 V_m) + i(\partial_x \mu_0 V_m + \partial_y A_z).$$

The expressions in parentheses are the Cauchy-Riemann equations, which are thus compactly incorporated into the expression [4]

$$\frac{\partial W}{\partial z^*} = 0. \tag{5.12}$$

The ∇ operator can be written as

$$\nabla = 2\frac{\partial}{\partial z^*}$$

$$\nabla^* = 2\frac{\partial}{\partial z} \tag{5.13}$$

and the Laplacian is

$$\nabla^2 = 4\frac{\partial^2}{\partial z \partial z^*}. \tag{5.14}$$

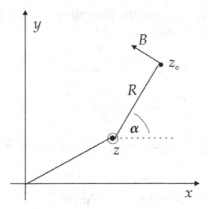

Figure 5.1 Geometry of a line current.

5.3 Field from a line current

Consider a single filament of current that crosses the x-y plane at the location z and an observation point z_o, as shown in Figure 5.1. The displacement between these two points is

$$z_o - z = Re^{ia}. \tag{5.15}$$

We know from Equation 1.26 that the field of the line current is

$$\vec{B} = \frac{\mu_0 I}{2\pi R}(-\hat{x}\sin\alpha + \hat{y}\cos\alpha). \tag{5.16}$$

Every filament in a magnet must have a return filament of the opposite polarity somewhere. It is convenient to assume that all filaments have their currents return through a filament at the coordinate origin. Considering Equation 3.9 for the vector potential of a line current, we assume the complex potential for a filament and its return is given by

$$W = -\frac{\mu_0 I}{2\pi}[\ln(z_o - z) - \ln(z_o - 0)].$$

The second term in this equation is constant for a given field point. In the cross-section of any real magnet, there are equal numbers of filaments with positive and negative currents. Thus summed over all the filaments in a magnet, the second terms cancel out. The first terms do not cancel out in general because the filaments have different positions z. Thus the potential for the line current is

$$W(z_o) = -\frac{\mu_0 I}{2\pi}\ln(z_o - z). \tag{5.17}$$

Using Equations 5.5 and 5.15 with the derivative operating on the coordinates of the field point z_o, we get the magnetic field

$$
\begin{aligned}
B_x - i\,B_y &= -i\,\frac{\mu_0 I}{2\pi}\,\frac{1}{z_o - z} \\[6pt]
&= -i\,\frac{\mu_0 I}{2\pi R}\,e^{-i\alpha} \\[6pt]
&= -i\,\frac{\mu_0 I}{2\pi R}\,(\cos\alpha - i\sin\alpha),
\end{aligned}
\tag{5.18}
$$

which agrees with the result in Equation 5.16.

From Equation 5.15, the complex logarithm is

$$
\ln(z_o - z) = \ln R + i\alpha.
$$

Using Equations 5.3 and 5.17 we can confirm that the vector potential for a line current is

$$
A_z = -\frac{\mu_0 I}{2\pi}\ln R
\tag{5.19}
$$

and find that the scalar potential is

$$
\begin{aligned}
V_m &= -\frac{I}{2\pi}\alpha \\[6pt]
&= -\frac{I}{2\pi}\tan^{-1}\!\left(\frac{y_o - y}{x_o - x}\right).
\end{aligned}
\tag{5.20}
$$

Although Equations 5.19 and 5.20 appear very different, they both lead to the fields in Equation 5.18.

Assume that a line current is located at position z. Then, according to Equation 5.18, the field intensity at the observation point z_o is

$$
H^*(z_o) = -i\,\frac{I}{2\pi}\,\frac{1}{z_o - z}.
\tag{5.21}
$$

Integrate H^* over observation points around any closed contour that encloses the point z.

$$
\oint H^*\,dz_o = -i\,\frac{I}{2\pi}\oint\frac{1}{z_o - z}\,dz_o.
$$

According to Cauchy's integral formula,[3]

---

[3] See Appendix E.

$$\oint \frac{1}{z_o - z} dz_o = 2\pi i.$$

Thus we find that the complex form of the Ampère law is

$$\oint H^* dz_o = I, \tag{5.22}$$

where I is the total current enclosed by the contour.

Now consider a line current in the vicinity of a plane surface of infinite permeability iron. As we saw in Chapter 2, the effect of the iron on the field of a conductor filament is equivalent to the presence of an image filament on the other side of the iron surface. The direction of the image current is the same as the conductor current. In the case of a circular boundary of radius R, the positions of the conductor and image filaments are

$$z = \rho e^{i\phi}$$
$$z_I = \frac{R^2}{\rho} e^{i\phi} = \frac{R^2}{z^*}. \tag{5.23}$$

5.4 Field from a current sheet

We can consider a current sheet as a collection of parallel line currents. The sheet is assumed to have a finite width, but to have infinitesimal thickness. Then using Equation 5.17, the potential for the current sheet is

$$W(z_o) = -\frac{\mu_0}{2\pi} \int K(s) \ln[z_o - z(s)] ds, \tag{5.24}$$

where s is the arc length along the sheet and $K = dI/ds$ is the sheet current density.

Example 5.1: potential for a straight sheet with constant K
Consider the straight sheet with width b shown in Figure 5.2. The current density is $K = \frac{I}{b}$ and the filaments making up the sheet are located at

$$z(s) = z_1 + se^{i\theta}.$$

It follows that $dz = e^{i\theta} ds$ and

$$z_2 = z_1 + be^{i\theta}. \tag{5.25}$$

From Equation 5.24, the potential is

$$W(z_o) = -\frac{\mu_0 I}{2\pi b} \int \ln[z_o - z] e^{-i\theta} dz.$$

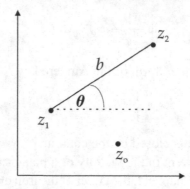

Figure 5.2 A straight current sheet.

We can write

$$\ln(z_o - z) = \ln(z - z_o) + \ln(-1)$$
$$= \ln(z - z_o) + i\pi.$$

The second term is independent of z_o and gets absorbed into the constant term for the potential. Defining $u = z - z_o$, we get

$$W(z_o) = -\frac{\mu_0 I}{2\pi b} e^{-i\theta} \int_{u_1}^{u_2} \ln u \; du.$$

Substituting for $e^{-i\theta}$ from Equation 5.25 and evaluating the integral[4] gives

$$W(z_o) = -\frac{\mu_0 I}{2\pi(z_2 - z_1)} [u_2 \ln u_2 - u_1 \ln u_1 - u_2 + u_1].$$

The last two terms in the square bracket give

$$-u_2 + u_1 = -(z_2 - z_o) + (z_1 - z_o)$$
$$= z_1 - z_2.$$

This term is also independent of z_o and gets absorbed into the constant term for the potential. Thus the potential for the straight sheet is given by [5]

$$W(z_o) = -\frac{\mu_0 I}{2\pi(z_2 - z_1)} [u_2 \ln u_2 - u_1 \ln u_1]. \tag{5.26}$$

Summing up the contributions to the magnetic field from the field of individual line currents given in Equation 5.21, we find the field of a current sheet is given by

---

[4] CRC 377.

$$H^*(z_o) = \frac{i}{2\pi} \int \frac{K(s)}{z(s) - z_o} ds, \tag{5.27}$$

where s is the distance along the sheet. If the position along the sheet is specified by the polar angle ϕ, this can be written as

$$H^*(z_o) = \frac{i}{2\pi} \int \frac{dI/d\phi}{z(\phi) - z_o} d\phi. \tag{5.28}$$

It is possible to determine a unique current distribution $dI/d\phi$ for circular or elliptic current sheets that can produce any desired two-dimensional field compatible with Maxwell's equations in the magnet aperture.[6]

Example 5.2: field due to a circular arc sheet with constant current density
Let us consider a current sheet in the form of a circular arc, as shown in Figure 5.3. The magnetic field from the sheet is given by Equation 5.28.

$$H^*(z_o) = -\frac{i}{2\pi} \frac{dI}{d\phi} \int_{\phi_1}^{\phi_2} \frac{d\phi}{z_o - z(\phi)}.$$

Since $z = ae^{i\phi}$, we can write this as

$$H^*(z_o) = -\frac{1}{2\pi} \frac{dI}{d\phi} \mathbb{I},$$

where[5]

$$\mathbb{I} = \int_{z_1}^{z_2} \frac{dz}{z(z_o - z)}$$

$$= \frac{2i}{z_o} \tan^{-1}\left[i\left(\frac{z_o - 2z}{z_o}\right)\right]. \tag{5.29}$$

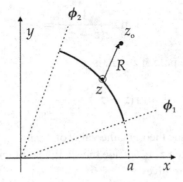

Figure 5.3 Circular arc current sheet.

[5] GR 2.172.

Therefore the field is

$$H^*(z_o) = -\frac{i}{\pi z_o}\frac{dI}{d\phi}\left\{\tan^{-1}\left[i\left(\frac{z_o - 2z}{z_o}\right)\right]\right\}_{z_1}^{z_2}.$$

Using the relation[6]

$$\tan^{-1}z = \frac{1}{2i}\ln\left(\frac{1 + iz}{1 - iz}\right),$$

we can write the field of the circular arc as

$$H^*(z_o) = -\frac{i}{2\pi z_o}\frac{dI}{d\phi}\left[\ln\left(\frac{z_2}{z_o - z_2}\right) - \ln\left(\frac{z_1}{z_o - z_1}\right)\right]. \qquad (5.30)$$

From Equation 5.28, the field of the arc conductor at the origin is

$$H^*(0) = \frac{i}{2\pi}\frac{dI}{d\phi}\int_{\phi_1}^{\phi_2}\frac{d\phi}{a\,e^{i\phi}}$$

$$= -\frac{1}{2\pi a}\frac{dI}{d\phi}[e^{-i\phi_2} - e^{-i\phi_1}].$$

If the angular arc completes a *full* circle, we have a current shell and the integral in Equation 5.29 becomes

$$\mathbb{I} = -\oint\frac{dz}{z(z - z_o)}.$$

A point where the denominator of the integrand becomes zero is called a *pole*. If z_o is inside the circle, then the contour integral has simple poles at $z = 0$ and $z = z_o$. The *residue*[7] for the pole at $z = 0$ is

$$\lim_{z\to 0} z\frac{1}{z(z - z_o)} = -\frac{1}{z_o}$$

and the residue for the pole at $z = z_o$ is

$$\lim_{z\to z_o}(z - z_o)\frac{1}{z(z - z_o)} = \frac{1}{z_o}.$$

Therefore, by the residue theorem, the value of the integral is zero and the field inside the shell vanishes. When z_o is outside the shell, the integral only has the pole at $z = 0$ and the residue theorem gives

[6] GR 1.622.3.   [7] See Appendix E.

$$\mathbb{I} = -2\pi i \left(-\frac{1}{z_o} \right) = \frac{2\pi i}{z_o}.$$

Therefore, the field outside the shell is

$$H^*(z_o) = -\frac{i}{z_o}\frac{dI}{d\phi}.$$

Since the total current is

$$I = 2\pi \frac{dI}{d\phi},$$

the field can be written as

$$H^*(z_o) = -i\frac{I}{2\pi z_o}.$$

This is the same as Equation 5.21 for the field of a line current located at the center of the shell.

Let us apply the Ampère law, Equation 5.22, for an infinitesimal rectangular contour across a current sheet, as shown in Figure 5.4. Then we have,

$$H_1^*(z_o)dz - H_2^*(z_o)dz = dI,$$

where dI is the current enclosed in the contour. In the limit where the distance perpendicular to the sheet approaches 0, the path of the observation points approach the path along the sheet and this results in the "current sheet theorem."[7]

$$H_1^*(z) - H_2^*(z) = \frac{dI}{dz}. \tag{5.31}$$

In addition to determining fields, this result has been used for calculations of magnetic stored energy and Lorentz forces.[8]

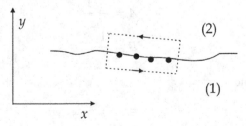

Figure 5.4 The sheet theorem.

5.5 cos ϕ current sheets

We consider here two examples of using complex methods to study the properties of current sheets that have a cos ϕ azimuthal current distribution.

Example 5.3:   field from cos ϕ current distribution using contour integration
Consider a closed circular sheet of radius a. We have

$$\frac{dI}{d\phi} = I_0 \cos \phi$$

$$z = ae^{i\phi}$$

$$\cos \phi = \frac{e^{i\phi} + e^{-i\phi}}{2} = \frac{z + z^*}{2a}.$$

Substituting into Equation 5.28, we get

$$
\begin{aligned}
H^*(z_o) &= \frac{i\,I_0}{4\pi a} \oint \frac{z + z^*}{z - z_o} \frac{dz}{i\,z} \\
&= \frac{I_0}{4\pi a} \oint \left[\frac{1}{z - z_o} + \frac{z^*}{z(z - z_o)} \right] dz \qquad\qquad (5.32) \\
&= \frac{I_0}{4\pi a} [\mathbb{I}_1 + \mathbb{I}_2].
\end{aligned}
$$

It follows from Cauchy's theorem that the first integral

$$\mathbb{I}_1 = \oint \frac{dz}{z - z_o} = \begin{cases} 2\pi i & \text{if } z_o < a \\ 0 & \text{if } z_o > a \end{cases}.$$

Using the method of partial fractions,[9] the denominator of the second integral can be written

$$\frac{1}{z(z - z_0)} = \frac{A}{z} + \frac{B}{z - z_o}$$
$$1 = (z - z_o)A + zB.$$

Equating powers of z, we find that

$$A = -\frac{1}{z_o}$$
$$B = \frac{1}{z_o}.$$

Then we can write

$$\mathbb{I}_2 = -\mathbb{I}_3 + \mathbb{I}_4,$$

where

$$\mathbb{I}_3 = \frac{1}{z_o} \oint \frac{z^*}{z} \, dz$$

$$= \frac{ia}{z_o} \int_0^{2\pi} e^{-i\phi} \, d\phi = 0$$

and

$$\mathbb{I}_4 = \frac{1}{z_o} \oint \frac{z^*}{z - z_o} \, dz. \qquad (5.33)$$

Using $z^* = a^2/z$, we can write this as

$$\mathbb{I}_4 = \frac{a^2}{z_o} \oint \frac{dz}{z(z - z_o)}$$

$$= \frac{a^2}{z_o} \left[-\frac{1}{z_o} \oint \frac{dz}{z} + \frac{1}{z_o} \oint \frac{dz}{z - z_o} \right] .$$

For z_o inside the contour, the factor in square brackets vanishes because of the residue theorem and $\mathbb{I}_4 = 0$. Then from Equation 5.32,

$$H^*(z_o) = H_x - iH_y$$

$$= \frac{I_0}{4\pi a} [2\pi i + 0] = i\frac{I_0}{2a} .$$

From this, we see that the field inside the current sheet is

$$H_x = 0$$

$$H_y = -\frac{I_0}{2a} . \qquad (5.34)$$

The field is only in the vertical direction and has constant magnitude everywhere inside the sheet in agreement with Equation 4.36.

For z_o outside the contour, we have[8]

$$\mathbb{I}_4 = \oint \frac{z^*}{z_o(z - z_o)} \, dz$$

$$= \frac{ia^2}{z_o} \int_0^{2\pi} \frac{1}{a \, e^{i\phi} - z_o} \, d\phi$$

$$= -\frac{a^2}{z_o^2} \, 2\pi i.$$

[8] GR 2.313.1.

Substituting these results back into Equation 5.32, the field outside the contour is

$$H^*(z_o) = \frac{I_0}{4\pi a}\left[0 - \frac{a^2}{z_o^2}2\pi i\right] = -i\frac{I_0 a}{2z_o^2}.$$

If we write $z_o = x + iy$ and multiply the numerator and denominator by $(z_o^*)^2$, we find that

$$H^*(z_o) = -\frac{I_0 a}{r^4}xy - i\frac{I_0 a}{2r^4}(x^2 - y^2).$$

Equating real and imaginary parts, we find the Cartesian field components outside the sheet are

$$
\begin{aligned}
H_x &= -\frac{I_0 a}{r^4}xy = -\frac{I_0 a}{2r^2}\sin 2\phi \\
H_y &= \frac{I_0 a}{2r^4}(x^2 - y^2) = \frac{I_0 a}{2r^2}\cos 2\phi.
\end{aligned}
\tag{5.35}
$$

On the midplane ($y = 0$), H is positive, along the y direction, and falls off with distance like $1/x^2$.

Example 5.4:   field from $\cos\phi$ current distribution using the sheet theorem
Assume again that we have a circular sheet with radius a. The current elements are located at

$$
\begin{aligned}
z &= ae^{i\phi} \\
dz &= izd\phi,
\end{aligned}
$$

so we have

$$
\begin{aligned}
\frac{dI}{dz} &= \frac{dI}{d\phi}\frac{d\phi}{dz} = -i\frac{I_0}{z}\cos\phi \\
&= -i\frac{I_0}{z}\left(\frac{e^{i\phi} + e^{-i\phi}}{2}\right) \\
&= -i\frac{I_0}{z}\left(\frac{z}{a} + \frac{a}{z}\right).
\end{aligned}
$$

Using the current sheet theorem, Equation 5.31,

$$H_1^*(z) - H_2^*(z) = -i\left(\frac{I_0}{2a} + \frac{I_0 a}{2z^2}\right).$$

The field inside the sheet H_{in} must be finite at $z = 0$ and for current in the positive z direction in the first quadrant of the circle, the field must go in the negative y direction. Therefore we identify H_2 with H_{in} and get

$$-H_{in}^*(z) = -H_{in,x} + iH_{in,y} = -i\frac{I_0}{2a}.$$

Equating real and imaginary parts, we find that

$$H_{in,x} = 0$$
$$H_{in,y} = -\frac{I_0}{2a} \tag{5.36}$$

in agreement with Equation 5.34. We identify the field exterior to the current sheet H_{ext} with H_1 in the sheet theorem.

$$H_{ext,x} - i\,H_{ext,y} = -i\frac{I_0 a}{2z^2}$$

5.6 Green's theorems in the complex plane

So far we have examined the fields due to current filaments and current sheets. We next want to proceed to the case of conductors with finite cross-sectional areas. However, before doing that, we need to review some important theorems that allows us to replace two-dimensional integrations over the conductor surface with contour integrals around the boundary of the surface. Besides the practical importance of reducing computation times in numerical calculations, this allows us to make use of some powerful results from the theory of complex contour integration.

Recall from Equation 3.79 that Green's theorem in the plane is

$$\iint \left(\frac{\partial Q}{\partial x} - \frac{\partial P}{\partial y}\right) dx\,dy = \oint (P\,dx + Q\,dy),$$

where $P(x,y)$ and $Q(x,y)$ are continuous functions with continuous partial derivatives in a region R that is bounded by a curve C. Define the complex function

$$F(z, z^*) = P(x,y) + iQ(x,y).$$

Using Equation 5.9, the derivative of F can be written as

$$2\frac{\partial F}{\partial z^*} = \left(\frac{\partial P}{\partial x} - \frac{\partial Q}{\partial y}\right) + i\left(\frac{\partial P}{\partial y} + \frac{\partial Q}{\partial x}\right). \tag{5.37}$$

The closed integral of F around C is

$$\oint F\,dz = \oint (P\,dx - Q\,dy) + i\oint (Q\,dx + P\,dy).$$

Applying Green's theorem in the plane, we have

$$\oint F \, dz = i \iint \left[\left(\frac{\partial P}{\partial x} - \frac{\partial Q}{\partial y} \right) + i \left(\frac{\partial P}{\partial y} + \frac{\partial Q}{\partial x} \right) \right] dx \, dy.$$

Replacing the integrand on the right-hand side using Equation 5.37, we find the first complex Green's theorem.[1]

$$\int \frac{\partial F}{\partial z^*} dS = \frac{1}{2i} \oint F \, dz \qquad (5.38)$$

Following similar arguments, we have

$$2 \frac{\partial F}{\partial z} = \left(\frac{\partial P}{\partial x} + \frac{\partial Q}{\partial y} \right) + i \left(-\frac{\partial P}{\partial y} + \frac{\partial Q}{\partial x} \right)$$

and

$$\oint F \, dz^* = -i \iint \left[\left(\frac{\partial P}{\partial x} + \frac{\partial Q}{\partial y} \right) + i \left(-\frac{\partial P}{\partial y} + \frac{\partial Q}{\partial x} \right) \right] dx \, dy.$$

After substitution, we obtain the second complex Green's theorem.[1]

$$\int \frac{\partial F}{\partial z} dS = -\frac{1}{2i} \oint F \, dz^*. \qquad (5.39)$$

5.7 Field from a block conductor

We next want to consider the case of a block conductor, which we define as one with finite cross-sectional area. If we consider the conductor block as made up from an array of current filaments, we can use Equation 5.21 and express the field as

$$H^* = -\frac{i}{2\pi} \int \frac{\sigma}{z_o - z} dS, \qquad (5.40)$$

where σ is the current density in the block. If we assume the current density is constant, we can rewrite this as

$$H^* = \frac{i\sigma}{2\pi} \int \frac{dS}{z - z_o}. \qquad (5.41)$$

Powerful methods have been developed that allow the fields from block conductors to be evaluated using contour integration.[1] Consider the Green's theorem, Equation 5.38. For our application, the integrand of the surface integral is associated with the expression for the magnetic field in Equation 5.41. The integral has

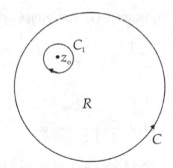

Figure 5.5 Contour for the Green's theorem calculation.

a singularity for the case when z_o is inside C, as shown in Figure 5.5. We can isolate the singularity by constructing a small circular contour C_1 around it. Then we can use Green's theorem in the region between the two contours to transform the surface integral into a contour integration. However, this requires evaluation on both contours C and C_1. Halbach proposed adding a constant term to the function F in Green's theorem that is (a) analytic in R and (b) makes the contour integration around C_1 vanish.[1] He assumed that F could be written as the product of two functions F_1 and F_2 with the property

$$\frac{\partial F}{\partial z^*} = F_1(z)\frac{\partial F_2(z^*)}{\partial z^*}, \tag{5.42}$$

where F_1 contains the singularity. Then F must have the form

$$F(z, z^*) = F_1(z)[F_2(z^*) - F_2(z_o^*)]. \tag{5.43}$$

Following this procedure, we define

$$F_1(z) = \frac{i\sigma}{2\pi}\frac{1}{z - z_o}$$
$$F_2(z^*) = z^*.$$

Then for use in Green's theorem, we have

$$F = \frac{i\sigma}{2\pi}\frac{1}{z - z_o}(z^* - z_o^*)$$

and

$$\frac{\partial F}{\partial z^*} = \frac{i\sigma}{2\pi}\frac{1}{z - z_o},$$

which is the integrand from Equation 5.41. Applying Green's theorem, we find that

$$H^* = \frac{1}{2i} \oint \frac{i\sigma}{2\pi} \frac{1}{z - z_o} (z^* - z_o^*) dz,$$

which simplifies to [1, 10]

$$H^* = \frac{\sigma}{4\pi} \oint \frac{z^* - z_o^*}{z - z_o} dz. \qquad (5.44)$$

Let us confirm that Equation 5.44 does indeed vanish for the circular contour C_1. Let

$$z - z_o = re^{i\theta}.$$

Then for the contour C_1,

$$\oint \frac{z^* - z_o^*}{z - z_o} dz = ir \int_0^{2\pi} e^{-i\theta} d\theta = 0.$$

Thus we can ignore the contours around isolated singularities inside the conductor region and only evaluate Equation 5.44 on the outer boundary of the conductor.

Other quantities of interest can also be conveniently expressed in terms of contour integrals. For example, the area A of a current block is given by [10, 11]

$$A = \frac{1}{2i} \oint z^* dz. \qquad (5.45)$$

Expressions have also been derived for the stored energy.[1, 12]

5.8 Block conductor examples

We consider three examples of using Equation 5.44 to find the field of a block conductor. The first example, the cylindrical conductor, was treated already in Chapter 1 using the Ampère law. Even though the calculation presented here is considerably more complicated, we carry it out to demonstrate some of the techniques involved and to compare with a result where we know the answer. The other two examples cannot be computed straightforwardly using the Ampère law.

Example 5.5:   field of a solid cylindrical conductor
Assume we have a solid cylindrical conductor with radius a, as shown in Figure 5.6. Let

$$z = ae^{i\phi}$$
$$z_o = re^{i\theta}.$$

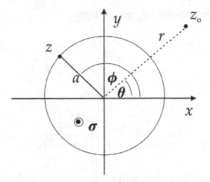

Figure 5.6 Solid cylindrical conductor.

Then Equation 5.44 gives

$$H^* = \frac{\sigma}{4\pi} \int_0^{2\pi} \frac{ae^{-i\phi} - re^{-i\theta}}{ae^{i\phi} - re^{i\theta}} \, i\, ae^{i\phi} \, d\phi$$

$$= i\, \frac{\sigma a}{4\pi} [a\, \mathbb{I}_1 - re^{-i\theta}\, \mathbb{I}_2], \tag{5.46}$$

where

$$\mathbb{I}_1 = \int_0^{2\pi} \frac{d\phi}{ae^{i\phi} - re^{i\theta}}$$

$$\mathbb{I}_2 = \int_0^{2\pi} \frac{e^{i\phi}}{ae^{i\phi} - re^{i\theta}} \, d\phi. \tag{5.47}$$

Case 1: z_o inside the conductor

When z_o is inside the conductor, we can use theorems from complex analysis to evaluate the integrals. Define $u = e^{i\phi}$ and $\beta = z_o/a$. Then

$$\mathbb{I}_1 = \frac{1}{ia} \oint \frac{du}{(u - \beta)u}. \tag{5.48}$$

We would like to convert the denominator into a simple pole so that we can use the residue theorem to evaluate the integral. To do this, expand the denominator using the method of partial fractions. Then we can write Equation 5.48 as

$$\mathbb{I}_1 = \frac{1}{ia} \left\{ -\frac{1}{\beta} \oint \frac{du}{u - 0} + \frac{1}{\beta} \oint \frac{du}{u - \beta} \right\}.$$

Applying the residue theorem, we obtain

$$\mathbb{I}_1 = \frac{1}{ia} 2\pi i \left(-\frac{1}{\beta} + \frac{1}{\beta} \right) = 0.$$

Turning next to $\mathbb{I}_2$,

$$\mathbb{I}_2 = \frac{1}{ia} \oint \frac{du}{(u - \beta)}$$

we apply the residue theorem and find

$$\mathbb{I}_2 = \frac{1}{ia} 2\pi i \, (1) = \frac{2\pi}{a}.$$

Returning now to Equation 5.46,

$$H^* = i \frac{\sigma a}{4\pi} (-r \, e^{-i\theta}) \frac{2\pi}{a}$$

$$= -i \frac{\sigma r}{2} [\cos \theta - i \sin \theta].$$

Thus the field inside the conductor is

$$H_x = -\frac{\sigma r}{2} \sin \theta$$

$$H_y = \frac{\sigma r}{2} \cos \theta. \tag{5.49}$$

Case 2: z_o outside the conductor

In this case, there are no singularities inside the contour, so we can treat $\mathbb{I}_1$ and $\mathbb{I}_2$ as ordinary integrals. Performing the first integration gives[9]

$$\mathbb{I}_1 = \frac{1}{-ire^{i\theta}} \left[i\phi - \ln(-re^{i\theta} + ae^{i\phi}) \right]_0^{2\pi}.$$

The logarithm term cancels because it has the same value at 0 and 2π. Thus we find that

$$\mathbb{I}_1 = -\frac{2\pi}{re^{i\theta}}.$$

---

[9] GR 2.313.1.

For the integral $\mathbb{I}_2$, let $\beta = r/a$ and $u = e^{i\phi}$. Then

$$\mathbb{I}_2 = \frac{1}{ia} \oint \frac{du}{u - \beta e^{i\theta}}$$

$$= \frac{1}{ia} \ln\left[e^{i\phi} - \beta e^{i\theta}\right]_0^{2\pi}.$$

The second term in the logarithm is constant and the first term has the same value at the two limits. Therefore, $\mathbb{I}_2 = 0$, and Equation 5.46 gives

$$H^* = -i\frac{\sigma a^2}{2re^{i\theta}}$$

$$= -i\frac{\sigma a^2}{2r}[\cos\theta - i\sin\theta].$$

Equating real and imaginary parts, we find the field outside the conductor is

$$H_x = -\frac{\sigma a^2}{2r}\sin\theta$$

$$\tag{5.50}$$

$$H_y = \frac{\sigma a^2}{2r}\cos\theta.$$

The field of an elliptical block conductor has also been found using similar methods.[12, 13]

Example 5.6: field outside a rectangular conductor

Assume we have a rectangular conductor oriented at an angle θ with respect to the x axis, as shown in Figure 5.7. We look for the field at the observation point z_o. In terms of the variables (z, z^*), a straight line from the vertex n to vertex $n + 1$ has the equation [1, 10]

$$z^* = z_n^* + \Delta z^*\left(\frac{z - z_n}{\Delta z}\right), \tag{5.51}$$

where

$$\Delta z = z_{n+1} - z_n. \tag{5.52}$$

Since the rectangle has four sides and has to close, we identify $z_5 = z_1$. For the side beginning with vertex 1, we define

$$\beta_1 = \frac{\Delta z^*}{\Delta z} = \frac{ae^{-i\theta}}{ae^{i\theta}} = e^{-2i\theta},$$

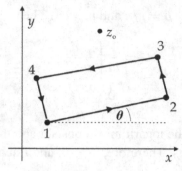

Figure 5.7 Rectangular conductor.

where a is the length of side 1. The second equation comes from considering z_1 as the origin of a line to z_2 in polar coordinates. Note that β is a constant because it is defined in terms of fixed z locations. In a rectangle, the angles of the other sides with respect to the x axis increase by 90° at each of the vertices. Thus for the side from vertex 2 to vertex 3,

$$\beta_2 = e^{-2i(\theta+\pi/2)} = e^{-i\pi}e^{-2i\theta} = -\beta_1.$$

Similarly we find,

$$\beta_3 = \beta_1$$
$$\beta_4 = -\beta_1.$$

Then Equation 5.44 gives

$$H^* = \frac{\sigma}{4\pi}\int_{z_1}^{z_2}\frac{[z_1^* + \beta_1(z-z_1) - z_o^*]}{z - z_o}\,dz + \cdots$$

with similar expressions for the remaining three sides. Defining the constant

$$\alpha_n = z_n^* - \beta_n z_n - z_o^*,$$

we can write

$$H^* = \frac{\sigma}{4\pi}[\alpha_1\mathbb{I}_1 + \beta_1\mathbb{I}_2] + \cdots. \qquad (5.53)$$

For points z_o outside the contour,

$$\mathbb{I}_1 = \int_{z_1}^{z_2}\frac{dz}{z - z_o}$$
$$= [\ln(z - z_o)]_{z_1}^{z_2}$$

and[10]

$$I_2 = \int_{z_1}^{z_2} \frac{z}{z - z_o} \, dz$$
$$= [z + z_o \ln(z - z_o)]_{z_1}^{z_2}.$$

Substituting into Equation 5.53,

$$H^* = \frac{\sigma}{4\pi} \left\{ \alpha_1 \ln\left(\frac{z_2 - z_o}{z_1 - z_o}\right) + \beta_1[z_2 + z_o\ln(z_2 - z_o) - z_1 - z_o\ln(z_1 - z_o)] \right\} + \cdots$$
$$= \frac{\sigma}{4\pi} \left\{ \alpha_1 \ln\left(\frac{z_2 - z_o}{z_1 - z_o}\right) + \beta_1 z_o \ln\left(\frac{z_2 - z_o}{z_1 - z_o}\right) + \beta_1(z_2 - z_1) \right\} + \cdots$$
$$= \frac{\sigma}{4\pi} \left\{ [\alpha_1 + \beta_1 z_o] \ln\left(\frac{z_2 - z_o}{z_1 - z_o}\right) + \beta_1(z_2 - z_1) \right\} + \cdots.$$

Writing out the third term for all four sides gives

$$\beta_1[(z_2 - z_1) - (z_3 - z_2) + (z_4 - z_3) - (z_1 - z_4)] = 2\beta_1[-z_1 + z_2 - z_3 + z_4].$$

For a rectangle, the directed line segments

$$z_4 - z_3 = -(z_2 - z_1),$$

so this term cancels. Thus we find the field at z_o due to the rectangular conductor block is [5, 10]

$$H^* = \frac{\sigma}{4\pi} \sum_{n=1}^{4} h_n$$
$$h_n = [(z_n - z_o)^* - \beta_n(z_n - z_o)] \ln\left(\frac{z_{n+1} - z_o}{z_n - z_o}\right). \tag{5.54}$$

Example 5.7: on-axis field for annular sector conductor

For our last example, consider a conductor with the shape of an annular sector, as shown in Figure 4.11. We look for the field at the center of the circular arcs. Thus we have

$$z = re^{i\phi}$$
$$z_o = 0$$

and the contour in Equation 5.44 can be broken into the four parts.

---

[10] GR 2.112.1.

$$H^*(0) = \frac{\sigma}{4\pi}\left\{\int_{r_1}^{r_2}\frac{re^{-i\phi_1}}{re^{i\phi_1}}e^{i\phi_1}dr + \int_{\phi_1}^{\phi_2}\frac{r_2e^{-i\phi}}{r_2e^{i\phi}}r_2i\,e^{i\phi}d\phi\right.$$

$$\left.+\int_{r_2}^{r_1}\frac{re^{-i\phi_2}}{re^{i\phi_2}}e^{i\phi_2}dr + \int_{\phi_2}^{\phi_1}\frac{r_1e^{-i\phi}}{r_1e^{i\phi}}r_1i\,e^{i\phi}d\phi\right\}.$$

Simplifying and performing the integrals, we get

$$H^*(0) = \frac{\sigma}{4\pi}\left\{\int_{r_1}^{r_2}e^{-i\phi_1}dr + i\int_{\phi_1}^{\phi_2}r_2e^{-i\phi}d\phi + \int_{r_2}^{r_1}e^{-i\phi_2}dr + i\int_{\phi_2}^{\phi_1}r_1e^{-i\phi}d\phi\right\}$$

$$= -\frac{\sigma}{2\pi}(r_2 - r_1)(e^{-i\phi_2} - e^{-i\phi_1}).$$

Expanding the exponentials, we find that the field of the annular sector conductor is

$$H^*(0) = -\frac{\sigma}{2\pi}(r_2 - r_1)[(\cos\phi_2 - \cos\phi_1) - i(\sin\phi_2 - \sin\phi_1)], \quad (5.55)$$

which agrees with Equation 4.50. Note that the field strength is proportional to the radial thickness.

5.9 Field from image currents

We now consider the magnetic field produced by a current distribution in the presence of infinite permeability iron. The case of a filament near a planar iron surface is shown in Figure 5.8. The current in the filament induces image currents on the surface of the iron, which can be represented by an equivalent image filament inside the iron. We have seen in Chapter 2 that the image current is in the same direction as the conductor filament and is located the same distance from the iron surface as the conductor filament. The field of the image filament is given by

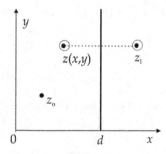

Figure 5.8 Line current near an iron slab.

$$H_I^*(z_o) = i\frac{I}{2\pi}\frac{1}{z_I - z_o}. \tag{5.56}$$

The location of the image filament is

$$z_I = d + (d - x) + i\,y$$
$$= 2d - z^*.$$

For a uniform distribution of current, we have

$$H_I^*(z_o) = i\frac{\sigma}{2\pi}\int\frac{dS}{2d - z^* - z_o}.$$

To convert this surface integral to a contour integral, we use the complex Green's theorem, Equation 5.39. Choosing

$$F = \frac{i\sigma}{2\pi}\frac{z}{2d - z^* - z_o},$$

we find that the contribution to the field from the image current in a planar iron surface is

$$H_I^*(z_o) = -\frac{\sigma}{4\pi}\oint\frac{z}{2d - z^* - z_o}dz^*. \tag{5.57}$$

We are also interested in the image currents near a circular iron surface at radius R, as shown in Figure 5.9. From Chapter 2, we know the image current is in the same direction as the conductor current. If ρ is the distance of the conductor filament from the center of the circle, then the image filament is a distance R^2/ρ from the center. Thus we have

$$z = \rho e^{i\phi}$$

$$z_I = \frac{R^2}{\rho}e^{i\phi} = \frac{R^2}{z^*}.$$

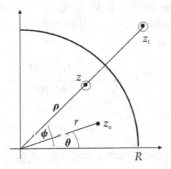

Figure 5.9 Line current near a circular iron cavity.

For a sheet conductor, the contribution of the images in a circular iron cavity to the field is a sum over the corresponding image currents. Using Equation 5.27, we obtain

$$
H_I^*(z_o) = \frac{i}{2\pi} \int \frac{K(z)}{z_I - z_o} dz
$$

$$
= \frac{i}{2\pi} \int \frac{Kz^*}{R^2 - z_o z^*} dz,
\tag{5.58}
$$

where $K(z) = dI/dz$.

Example 5.8: image field for $\cos\phi$ current sheet in an iron cavity

Let us examine the image field at the origin for a closed circular sheet with radius a and a $\cos\phi$ angular current distribution. When $z_o = 0$, the integrand does not have a singularity. Applying Equation 5.58,

$$
H_I^*(0) = \frac{i}{2\pi} \int_0^{2\pi} \frac{I_0 \cos\phi}{iz} \frac{ae^{-i\phi}}{R^2} iz d\phi
$$

$$
= \frac{iI_0 a}{2\pi R^2} \int_0^{2\pi} \frac{\cos\phi}{e^{i\phi}} d\phi
$$

$$
= \frac{iI_0 a}{4\pi R^2} \int_0^{2\pi} \frac{e^{i\phi} + e^{-i\phi}}{e^{i\phi}} d\phi.
$$

After the integration, we find the contribution of the image field at the origin is

$$
H_I^*(0) = i \frac{I_0 a}{2R^2} = H_{Ix} - iH_{Iy},
$$

which gives the field components

$$
H_{Ix} = 0
$$

$$
H_{Iy} = -\frac{I_0 a}{2R^2}.
\tag{5.59}
$$

The contribution of the image field is in the same direction that we saw in Equation 5.34 for the field of the conductor. The enhancement of the field due to the presence of the iron is

$$
E(0) = \frac{H^*(0) + H_I^*(0)}{H^*(0)}
$$

$$
= 1 + \frac{a^2}{R^2}.
\tag{5.60}
$$

This shows that the iron cavity can contribute up to a factor of 2 to the field at the origin.

For a block conductor with constant current density, the image current in circular iron is

$$H_I^*(z_o) = \frac{i\sigma}{2\pi} \int \frac{z^*}{R^2 - z_o z^*} \, dS.$$

Using Equation 5.39 for Green's theorem and choosing

$$F = \frac{i\sigma}{2\pi} \frac{zz^*}{R^2 - z_o z^*},$$

we find that the field due to the image current in the circular iron is [1]

$$H_I^*(z_o) = -\frac{\sigma}{4\pi} \oint \frac{zz^*}{R^2 - z_o z^*} \, dz^*. \tag{5.61}$$

Example 5.9: on-axis image field in circular iron for annular sector conductor
Consider an annular sector conductor extending from radius r_1 to r_2 inside an iron cavity of radius R. The image field at the origin is given by Equation 5.61 with $z_o = 0$.

$$H_I^*(0) = -\frac{\sigma}{4\pi R^2} \oint zz^* \, dz^*$$

$$= -\frac{\sigma}{4\pi R^2} \left\{ e^{-i\phi_1} \int_{r_1}^{r_2} r^2 \, dr - ir_2^3 \int_{\phi_1}^{\phi_2} e^{-i\phi} \, d\phi + e^{-i\phi_2} \int_{r_2}^{r_1} r^2 \, dr - ir_1^3 \int_{\phi_2}^{\phi_1} e^{-i\phi} \, d\phi \right\}.$$

Evaluating the integrals and simplifying gives the field contribution due to the iron.

$$H_I^*(0) = -\frac{\sigma}{6\pi R^2} (r_2^3 - r_1^3)[(\cos\phi_2 - \cos\phi_1) - i(\sin\phi_2 - \sin\phi_1)]. \tag{5.62}$$

Note that this expression has the same sign and angular dependence as the field from the conductor given in Equation 5.55. The presence of the iron gives the enhancement factor at the origin [14]

$$E(0) = 1 + \frac{r_2^2 + r_1 r_2 + r_1^2}{3R^2}. \tag{5.63}$$

5.10 Multipole expansion

Since the magnetic potential, Equation 5.3, is an analytic function, it can be expanded in a power series

$$W(z_o) = \sum_{n=0}^{\infty} w_n z_o^n.$$

The magnetic field can then be expressed as

$$H^*(z_o) = \frac{i}{\mu_0} \frac{dW}{dz_o}$$

$$= \sum_{n=1}^{\infty} \frac{i\,n}{\mu_0} w_n z_o^{n-1}.$$

Redefining the coefficients, we write the field as the power series

$$H^*(z_o) = \sum_{n=1}^{\infty} c_n z_o^{n-1}. \tag{5.64}$$

The field can also be expressed in terms of the integral in Equation 5.41.

$$H^*(z_o) = \frac{i}{2\pi} \int \frac{\sigma}{z - z_o}\, dS$$

$$= \frac{i}{2\pi} \int \frac{\sigma}{z\left(1 - \dfrac{z_o}{z}\right)}\, dS.$$

Expand the factor in the denominator in a geometric series.

$$H^*(z_o) = \frac{i}{2\pi} \int \frac{\sigma}{z}\left[1 + \frac{z_o}{z} + \left(\frac{z_o}{z}\right)^2 + \cdots\right] dS$$

This series converges for observation points inside the magnet aperture up to the closest conductor. Equating this expression with Equation 5.64 gives

$$\sum_{n=1}^{\infty} c_n z_o^{n-1} = \frac{i}{2\pi} \int \frac{\sigma}{z} \sum_{n=1}^{\infty} \left(\frac{z_o}{z}\right)^{n-1} dS$$

$$= \frac{i}{2\pi} \sum_{n=1}^{\infty} \int \frac{\sigma}{z} \left(\frac{z_o}{z}\right)^{n-1} dS.$$

The z_o factor cancels from both sides of the equation. Then matching term by term, we find

$$c_n = \frac{i\sigma}{2\pi} \int z^{-n} dS. \tag{5.65}$$

We can convert this surface integral into a contour integral by using the Green's theorem, Equation 5.39, with

$$F = \frac{i\sigma}{2\pi} \frac{z^{1-n}}{1-n}.$$

Thus Equation 5.65 becomes [1]

$$c_n = -\frac{1}{2i}\oint\frac{i\sigma}{2\pi}\frac{z^{1-n}}{1-n}dz^*$$

$$= \frac{\sigma}{4\pi(n-1)}\oint z^{1-n}dz^* \tag{5.66}$$

for $n > 1$. For the case $n = 1$, we return to Equation 5.65 and find

$$c_1 = \frac{i\sigma}{2\pi}\int\frac{1}{z}dS.$$

This time we use the Green's theorem Equation 5.38 with

$$F = \frac{i\sigma}{2\pi z}z^*$$

to find that [1]

$$c_1 = \frac{\sigma}{4\pi}\oint\frac{z^*}{z}dz. \tag{5.67}$$

Example 5.10: multipoles for an annular sector conductor

We consider an annular sector conductor with radius between r_1 and r_2 that has constant current density σ. Let $z = re^{i\phi}$. For multipoles with $n > 1$, we have using Equation 5.66

$$c_n = \frac{\sigma}{4\pi(n-1)}\left\{\int_{r_1}^{r_2}(re^{i\phi_1})^{1-n}e^{-i\phi_1}dr - i\int_{\phi_1}^{\phi_2}(r_2e^{i\phi})^{1-n}r_2e^{-i\phi}d\phi\right.$$

$$\left. + \int_{r_2}^{r_1}(re^{i\phi_2})^{1-n}e^{-i\phi_2}dr - i\int_{\phi_2}^{\phi_1}(r_1e^{i\phi})^{1-n}r_1e^{-i\phi}d\phi\right\}.$$

After performing the integrations and simplifying the algebraic results, we find that

$$c_n = -\frac{\sigma}{2\pi n(2-n)}(r_2^{2-n} - r_1^{2-n})(e^{-in\phi_2} - e^{-in\phi_1}). \tag{5.68}$$

Because of the factor in the denominator, this relation cannot be used when $n = 2$. For that case, we return to Equation 5.66 and find

$$c_2 = \frac{\sigma}{4\pi}\oint z^{-1}dz^*.$$

For the annular sector, performing the integrals and summing terms, we find the quadrupole multipole is

$$c_2 = -\frac{\sigma}{4\pi} \ln\left(\frac{r_2}{r_1}\right) \left(e^{-2i\phi_2} - e^{-2i\phi_1}\right). \qquad (5.69)$$

We can get the $n = 1$ term from Equation 5.67. The dipole multipole is

$$c_1 = -\frac{\sigma}{2\pi} (r_2 - r_1)\left(e^{-i\phi_2} - e^{-i\phi_1}\right). \qquad (5.70)$$

Errors in the construction of magnet coils can lead to the introduction of additional unwanted multipole contributions to the field.[1, 15] These errors can include left-right and up-down asymmetries in the shape of the coils, displacements, rotations, and errors in the excitation currents.

5.11 Field due to a magnetized body

We next look at the magnetic field produced by a magnetized body. This is the case, for example, for a permanent magnet with net magnetization in the x-y plane. Consider a pair of parallel filaments with currents flowing in opposite directions located a distance d apart, as shown in Figure 5.10. There is a net field component in the x-y plane, oriented perpendicular to the axis connecting the two filaments. Such an arrangement is known as a current *doublet*.[16, 17] The field for the two filaments is

$$H^*(z_o) = \frac{iI}{2\pi}\left[\frac{1}{z_2 - z_o} - \frac{1}{z_1 - z_o}\right].$$

Let

$$d = z_2 - z_1 = |d|e^{i\alpha}$$
$$z_d = \tfrac{1}{2}(z_1 + z_2),$$

where α is the angle between d and the x axis. Substituting, we find

$$H^*(z_o) = -\frac{iI}{2\pi}\left[\frac{d}{(z_d - z_o)^2 - \dfrac{d^2}{4}}\right].$$

Recall that the magnetic dipole moment is

$$m = IA = I\,l\,d,$$

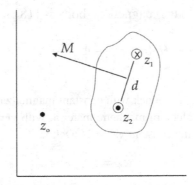

Figure 5.10 Model for a magnetized body.

where l is a unit distance along the z direction. Define m' as the magnetic moment per unit length. Then in the limit as $d \to 0$,

$$I\,d \to m'$$

$$\frac{d^2}{4} \to 0$$

and the field at z_o due to the doublet at z is

$$H^*(z_o) = -\frac{i}{2\pi} \frac{m'\,e^{i\alpha}}{(z - z_o)^2}.$$

We now want to express the field in terms of the magnetization M. The direction of m' is rotated by $\pi/2$ with respect to the direction of d. Let us define β to be the direction of M with respect to the x axis.

$$\beta = \alpha - \frac{\pi}{2}$$

$$e^{i\alpha} = e^{i\beta}\,e^{i\pi/2} = ie^{i\beta}.$$

Then summing up all the magnetic moments in the magnetized body, we have [18]

$$H^*(z_o) = \frac{1}{2\pi} \int \frac{M}{(z - z_o)^2}\,dS. \qquad (5.71)$$

If the magnetization is constant in the body, we can convert this to a contour integral by using the Green's theorem, Equation 5.39. Defining

$$F = -\frac{M}{2\pi} \frac{1}{z - z_o},$$

we find that the field due to the magnetized body is [18]

$$H^*(z_o) = \frac{M}{4\pi i} \oint \frac{dz^*}{z - z_o}.$$ (5.72)

Example 5.11:   triangular block with constant magnetization

Consider a triangular block of magnetic material with vertices $\{z_1, z_2, z_3\}$. Assume the magnetization has the constant value M. Define

$$\Delta z_n = z_{n+1} - z_n.$$

Since the triangle is closed, $z_4 = z_1$. The slopes of the sides are

$$\beta_n = \frac{\Delta z_n^*}{\Delta z_n},$$

so we can change the integration variable in Equation 5.72 from z^* to z for side n through the relation

$$dz^* = \beta_n dz.$$

The field produced by the block is

$$H^*(z_o) = \frac{M}{4\pi i} \left[\int_{z_1}^{z_2} \frac{\beta_1}{z - z_o} \, dz + \cdots \right]$$

$$= \frac{M}{4\pi i} [\beta_1 \ln(z_2 - z_o) - \beta_1 \ln(z_1 - z_o) + \cdots].$$

Collecting terms, the field of the triangular magnetized block is

$$H^*(z_o) = \frac{M}{4\pi i} \{ (\beta_3 - \beta_1)\ln(z_1 - z_o) + (\beta_1 - \beta_2)\ln(z_2 - z_o)$$

$$+ (\beta_2 - \beta_3)\ln(z_3 - z_o) \}.$$ (5.73)

5.12 Force

The vector force dF on a current filament is

$$\vec{dF} = I\vec{dl} \times \vec{B}.$$

If the filament is directed along the z direction, B is in the x-y plane, and so is F. The force can then be written as the complex variable $F = F_x + iF_y$, where

$$dF_x = -\mu_0 I H_y dz$$

$$dF_y = \mu_0 I H_x dz.$$

Define f to be the force per unit length in the z direction.

$$f = \frac{dF}{dz} = i\mu_0 \, I \, H$$

For a distributed current distribution, we can generalize this as

$$f = i\mu_0 \int \sigma \, H dS. \tag{5.74}$$

Using Equation 5.11 for σ, we have

$$f = 2\mu_0 \int H \frac{\partial H}{\partial z} dS.$$

To express this as a contour integral, use the complex Green's theorem, Equation 5.39, with

$$F = \mu_0 H^2,$$

which gives [1]

$$f = \frac{i\mu_0}{2} \oint H^2 dz^*. \tag{5.75}$$

This shows that the transverse force per unit length is proportional to the square of the magnetic field intensity. Examples of complex force calculations can be found in references.[1, 19]

5.13 Conformal mapping

Operating on a complex variable z with some function f

$$w = f(z)$$

produces another complex variable w. This can be interpreted as a mapping from the z plane onto another w plane. Suppose that two curves in the z plane intersect at a point with the angle θ between them. A mapping is called *conformal* if the two corresponding curves in the w plane also intersect with the same angle θ between them. If $f(z)$ is an analytic function with $df/dz \neq 0$ inside a region R, then the mapping is conformal. Conformal mappings have the property that the function in the w plane is also analytic, so the real and imaginary parts of the mapped function are solutions of the Laplace equation.

Conformal mapping can frequently be used to transform a problem with complicated boundaries in the z plane, for example, into a simpler problem in the upper half-plane or the interior of the unit circle in the w plane. Once the solution is found for the problem in the w plane, an inverse mapping $z = g(w)$ can be used to obtain the solution to the original problem. The theory of conformal mapping is a major subject in its own right. We only have space here to briefly introduce the subject and present a few examples. Fortunately, approximately half the book by Binns and Lawrenson is devoted to using conformal mapping in the solution of electric and magnetic field problems.[20] The interested reader can find many useful examples there.

The bilinear transformation combines the operations of translation, rotation, stretching, and inversion.[21]

$$w = \frac{\alpha z + \beta}{\gamma z + \delta},$$

where α, β, γ, and δ are complex numbers with the property that

$$\alpha\delta - \beta\gamma \neq 0.$$

This transformation can map circles and lines in the z plane into circles and lines in the w plane. It can be used, for example, to map a pair of separated circles to concentric circles. The bilinear transformation has the property that a quantity known as the cross-ratio is conserved.

$$\frac{(w - w_1)(w_2 - w_3)}{(w - w_3)(w_2 - w_1)} = \frac{(z - z_1)(z_2 - z_3)}{(z - z_3)(z_2 - z_1)} \tag{5.76}$$

This expression can be used to create a transformation that maps three given points in the z plane to three corresponding points in the w plane. An important bilinear transformation that maps any point z_o in the upper half of the z plane into the interior of the unit circle in the w plane is given by [22]

$$w = e^{i\theta_0} \left(\frac{z - z_o}{z - z_o^*} \right). \tag{5.77}$$

The points on the x axis are mapped to the boundary of the circle.

Example 5.12: line current in an iron cavity
Suppose we have a line current at the point w_1 inside a circular cavity with unit radius that is made from infinitely permeable iron, as shown in Figure 5.11. We use Equation 5.77 to map between the physical situation in the w plane and the upper half of the z plane. To determine the two unknown constants θ_0 and z_o, we associate the points

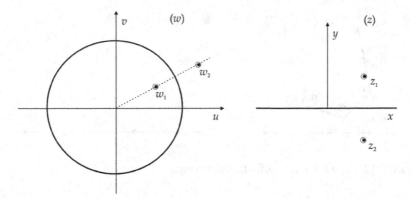

Figure 5.11 Line current in an iron cavity.

$$z = i \leftrightarrow w = 0$$
$$z = \infty \leftrightarrow w = -1,$$

which requires that

$$z_o = i$$
$$e^{i\theta_0} = -1.$$

This gives the specific mapping function between the planes

$$w = \frac{i - z}{i + z}.$$

The known line current at w_1 maps to a line current at z_1 in the z plane and the circular iron boundary maps to an iron plane along the real axis in the z plane. The mapping between the two line currents is

$$w_1 = u_1 + iv_1 = \frac{i - x_1 - iy_1}{i + x_1 + iy_1}.$$

Normalizing the denominator, we find

$$u_1 + iv_1 = \frac{[1 - x_1^2 - y_1^2] + i\,[2x_1]}{x_1^2 + (1 + y_1)^2}.$$

The real and imaginary parts of this equation can be solved for x_1 and y_1 as

$$x_1 + iy_1 = \frac{[2v_1] + i\,[1 - u_1^2 - v_1^2]}{u_1^2 + v_1^2 + 2u_1 + 1}.$$

In the z plane, we know there is an image current below the iron plane at the location $z_2 = z_1^*$. We can then use the mapping function to find the location w_2 of the image current in the w plane. After some algebraic simplifications, we find that

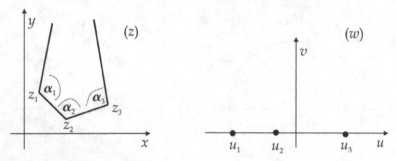

Figure 5.12 Schwarz-Christoffel transformation.

$$w_2 = u_2 + iv_2 = \frac{-v_1 + i(u_1 + 1)}{v_1 + i(u_1^2 + v_1^2 + u_1)}$$

$$= \frac{u_1 + iv_1}{u_1^2 + v_1^2}.$$

Representing w_1 and w_2 in polar coordinates, we find that

$$r_2 = \frac{1}{r_1}$$
$$\theta_2 = \theta_1,$$

which agrees with the result from the method of images.

Suppose that the boundary of some region in the z plane is made up of a series of straight line segments, as shown in Figure 5.12. The line segments meet at the vertices $z_1, z_2, \ldots$ It is possible to map this boundary to the real axis in the w plane by using the Schwarz-Christoffel transformation,[23] which takes the form of the differential equation

$$\frac{dz}{dw} = G(w - u_1)^{\alpha_1/\pi - 1}(w - u_2)^{\alpha_2/\pi - 1} \cdots (w - u_n)^{\alpha_n/\pi - 1}, \qquad (5.78)$$

where G is a complex constant and the α_i are the interior angles. The points $u_1, u_2, \ldots$ on the real axis of the w plane correspond to the vertices in the z plane. The interior of the figure in the z plane maps to the upper half of the w plane.

Example 5.13:   potential of a line current near the corner of two perpendicular planes
Consider a line current near the perpendicular intersection of two infinitely permeable plane surfaces, as shown in Figure 5.13. We solve the problem by using the Schwarz-Christoffel transformation, which in this case takes the form

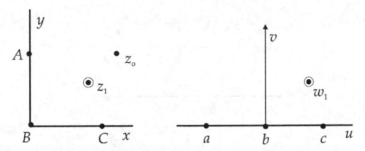

Figure 5.13 Line current near a corner. *ABC* and *abc* lie on infinitely permeable boundary surfaces. The line current is at z_1.

$$\frac{dz}{dw} = G(w - u_1)^{-1/2},$$

since the vertex angle $\alpha_1 = \pi/2$. Integrating this equation, we find

$$z = 2G\sqrt{w - u_1} + H,$$

where H is another complex constant. Solving for w, we get

$$w - u_1 = \frac{z^2 - 2Hz + H^2}{4G^2}.$$

Breaking this equation into real and imaginary parts, leads to

$$u - u_1 + iv = \frac{x^2 - y^2 - 2Hx + H^2 + 2i(xy - Hy)}{4G^2}. \tag{5.79}$$

We choose three points A, B, C on the boundary in the z plane and demand that they correspond to three points a, b, c along the real axis in the w plane according to the following prescription:

$$A : x = 0, y = 1 \leftrightarrow a : u = -1, v = 0$$
$$B : x = 0, y = 0 \leftrightarrow b : u = 0, v = 0$$
$$C : x = 1, y = 0 \leftrightarrow c : u = 1, v = 0$$

Applying these constraints to Equation 5.79, we find that

$$u_1 = 0$$
$$H = 0$$
$$4G^2 = 1$$

and the resulting transformation equation is

$$w = z^2.$$

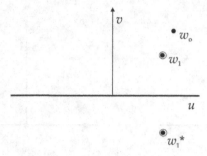

Figure 5.14 Image current in the *w* plane.

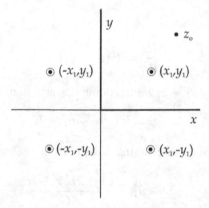

Figure 5.15 Line current and three images in the *z* plane.

In the *w* plane, the potential is due to the line current w_1 and the image current due to the plane boundary of the infinitely permeable material, as shown in Figure 5.14. The potential is given by

$$W(w_o) = \frac{\mu_0 I}{2\pi}[\ln(w_o - w_1) + \ln(w_o - w_1^*)].$$

Transforming back to the *z* plane, we have

$$W(z_o) = \frac{\mu_0 I}{2\pi}[\ln(z_o^2 - z_1^2) + \ln(z_o^2 - z_1^{*2})]$$

$$= \frac{\mu_0 I}{2\pi}\left\{\ln[(z_o - z_1)(z_o + z_1)] + \ln[(z_o - z_1^*)(z_o + z_1^*)]\right\}$$

$$= \frac{\mu_0 I}{2\pi}\left[\ln(z_o - z_1) + \ln(z_o + z_1) + \ln(z_o - z_1^*) + \ln(z_o + z_1^*)\right].$$

This shows that the potential in the *z* plane is due to the physical line current at z_1 together with three image currents,[24] as shown in Figure 5.15. The four currents lie on a circle centered at the corner of the iron surfaces.

5.14 Integrated potentials

Suppose that $\Phi(r, \theta, s)$ is a scalar potential describing some three-dimensional magnetic configuration. Assume the conductors have a finite extent in the s direction, so that Φ vanishes as $s \to \pm\infty$. Define

$$E(r, \theta) = \int_{-\infty}^{\infty} \Phi(r, \theta, s) \, ds.$$

Taking the derivative with respect to r gives

$$\frac{\partial E}{\partial r} = \int_{-\infty}^{\infty} \frac{\partial \Phi}{\partial r} \, ds = -\int_{-\infty}^{\infty} B_r \, ds, \tag{5.80}$$

while the derivative with respect to θ yields

$$\frac{1}{r} \frac{\partial E}{\partial \theta} = \int_{-\infty}^{\infty} \frac{1}{r} \frac{\partial \Phi}{\partial \theta} \, ds = -\int_{-\infty}^{\infty} B_\theta \, ds. \tag{5.81}$$

We can likewise define $A(r, \theta, s)$ as the vector potential describing the same three-dimensional magnetic configuration. Since the conductors have a finite extent in the s direction, A_s also vanishes as $s \to \pm\infty$. Define

$$F(r, \theta) = \int_{-\infty}^{\infty} A_s(r, \theta, s) \, ds.$$

Considering the integral of B_r, we find that

$$
\int_{-\infty}^{\infty} B_r ds = \int (\nabla \times \vec{A})_r ds
$$
$$
= \int \left(\frac{1}{r} \frac{\partial A_s}{\partial \theta} - \frac{\partial A_\theta}{\partial s} \right) ds
$$
$$
= \frac{1}{r} \frac{\partial}{\partial \theta} \int A_s ds - A_\theta \Big|_{-\infty}^{\infty}.
$$

Assuming that A_θ has the same value at $\pm\infty$, we find that

$$\int_{-\infty}^{\infty} B_r \, ds = \frac{1}{r} \frac{\partial F}{\partial \theta}. \tag{5.82}$$

Similarly, the integral of B_θ

$$
\int_{-\infty}^{\infty} B_\theta ds = \int \left(\frac{\partial A_r}{\partial s} - \frac{\partial A_s}{\partial r} \right) ds
$$
$$
= A_r \Big|_{-\infty}^{\infty} - \frac{\partial}{\partial r} \int A_s \, ds,
$$

so that

$$\int_{-\infty}^{\infty} B_\theta \, ds = -\frac{\partial F}{\partial r}. \tag{5.83}$$

Equating the expressions for the integrated values of B_r and B_θ, we find that

$$\frac{1}{r} \frac{\partial F}{\partial \theta} = -\frac{\partial E}{\partial r}$$

$$\frac{\partial F}{\partial r} = \frac{1}{r} \frac{\partial E}{\partial \theta}.$$

These two equations have the same form as the Cauchy-Riemann equations in polar coordinates. Thus F and E represent the real and imaginary parts of the analytic potential function

$$W(z) = F(r, \theta) + iE(r, \theta).$$

This potential can be used to describe the influence of a magnet end on the field quality of a long magnet.[25]

References

[1] K. Halbach, Fields and first order perturbation effects in two dimensional conductor dominated magnets, *Nuc. Instr. Meth.* 78:185, 1970.

[2] K. Miller, *Introduction to Advanced Complex Calculus*, Dover, 1970, p. 66.

[3] M. Spiegel, *Complex Variables, Schaum's Outline Series*, McGraw-Hill, 1964, p. 7, 69–70, 83.

[4] R. Beth, Complex methods for three-dimensional magnetic fields, *IEEE Trans. Nuc. Sci.* 18:901, 1971.

[5] R. Beth, Complex representation and computation of two-dimensional magnetic fields, *J. Appl. Phys.* 37:2568, 1966.

[6] R. Beth, Elliptical and circular current sheets to produce a prescribed internal field, *IEEE Trans. Nuc. Sci.* 14:386, 1967.

[7] R. Beth, Some extensions of complex methods for two-dimensional fields, *Proc. 6th Int. Conf. on High Energy Accelerators*, Cambridge Electron Accelerator Lab, Cambridge, MA, 1967, p. 387.

[8] F. Toral, et al., Further developments on Beth's current sheet theorem: computation of magnetic field, energy and mechanical stresses in the cross section of particle accelerator magnets, *IEEE Trans. Appl. Superconductivity* 14:1886, 2004.

[9] M. Protter & C. Morrey, *College Calculus with Analytic Geometry*, Addison-Wesley, 1964, p. 465–468.

[10] R. Beth, Evaluation of current produced two-dimensional magnetic fields, *J. Appl. Phys.* 40:4782, 1969.

[11] M. Spiegel, *op. cit.*, p. 114.

[12] R. Beth, Analytic design of superconducting multipolar magnets, Proc. 1968 *Summer Study on Superconducting Devices and Accelerators*, Brookhaven National Laboratory, p. 843–859.

[13] R. Beth, An integral formula for two-dimensional fields, *J. Appl. Phys.* 38:4689, 1967.

[14] J. Blewett, Iron shielding for air core magnets, Proc. 1968 *Summer Study on Superconducting Devices and Accelerators*, Brookhaven National Laboratory, p. 1042–1051.

[15] A. Jain, Basic theory of magnets, in S. Turner (ed.), *CERN Accelerator School on Measurement and Alignment of Accelerator and Detector Magnets*, CERN 98–05, 1998, p. 1.

[16] K. Binns & P. Lawrenson, *Analysis and Computation of Electric and Magnetic Field Problems*, 2nd ed., Pergamon Press, 1973, p. 48–49.

[17] M. Green, Modeling the behavior of oriented permanent magnet material using current doublet theory, *IEEE Trans. Mag.* 24:1528, 1988.

[18] K. Halbach, Design of permanent multipole magnets with oriented rare earth cobalt material, *Nuc. Instr. Meth.* 169:1, 1980.

[19] R. Beth, Currents and coil forces as contour integrals in two dimensional magnetic fields, *J. Appl. Phys.* 40:2445, 1969.

[20] K. Binns & P. Lawrenson, *op. cit.*, chapters 6–10 and appendix III.

[21] K. Miller, *op. cit.*, p. 202–204.

[22] M. Spiegel, *op. cit.*, p. 203–204, 216.

[23] J. Dettman, Applied Complex Variables, Dover, 1984, p. 260–265.

[24] K. Binns & P. Lawrenson, *op. cit.*, p. 43.

[25] F. Mills & G. Morgan, A flux theorem for the design of magnet coil ends, *Part. Acc.* 5:227, 1973.

6

Iron-dominant transverse fields

In the previous two chapters, we have been discussing transverse field magnets where the field shape is controlled by the distribution of the conductors. When iron was present, it served mainly to enhance the strength of the field produced by the conductors. In this chapter, we examine transverse field magnets where the primary roles of the conductor and the iron are reversed. Here the shape of the field is determined by the shape of the iron surface and the conductors are used to excite the field in the iron.[1, 2] In addition, the iron reduces the reluctance in the magnetic circuit, allowing a larger useful field for a given number of amp-turns from the conductor. These types of magnets typically have a maximum field less than 2 T, so that iron saturation effects do not destroy the field quality. We will mainly be concerned with the calculation of the magnetic fields and do not consider the many engineering considerations necessary to actually build magnets of this type.

6.1 Ideal multipole magnets

If the permeability of the iron is very large ($\mu_r \sim 1000$), it is a useful approximation to assume that μ_r is infinite. In that case, the magnetic flux density B must be perpendicular to the iron surface. The shape of the iron surface in the transverse plane coincides with an equipotential line for the scalar potential. Then, since the equipotential lines of the real and imaginary parts of the complex potential W are orthogonal, the magnetic field follows from the equipotential lines for the vector potential. Each positive pole of the magnet acts like a source of magnetic field, while the negative poles act like a sink where the magnetic field returns back into the iron.

The shape of the iron pole piece for an ideal $2n$-multipole magnet is determined by the complex potential for the multipole, which can be found from conformal mapping to have the form

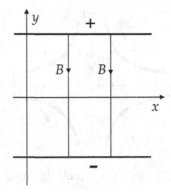

Figure 6.1 Iron surfaces and field lines in a dipole magnet.

$$W(z) = c_n z^n$$
$$= c_n\, r^n e^{in\theta},$$

where the constant c_n gives the strength of the potential. The magnetic field for the ideal multipole is given by

$$B^* = i\,\frac{dW}{dz}$$
$$= i\,n\,c_n z^{n-1}.$$

The simplest example of an ideal multipole is the dipole ($n = 1$), which has the complex potential

$$W = c_1 z$$
$$= c_1(x + i\,y).$$

Figure 6.1 illustrates the iron surface and the lines of magnetic field for a dipole magnet. The dipole has two poles with opposite polarity. Taking the real and imaginary parts of W, the vector and scalar potentials are

$$A_z = c_1\,x$$
$$\mu_0 V_m = c_1\,y.$$

We see that the iron surface is given by the equipotential

$$c_1\,y = h,$$

where the constant h identifies a particular surface. The vector potential is given by the equipotential

$$c_1\,x = k,$$

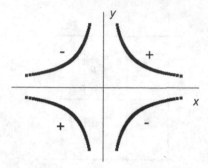

Figure 6.2 Iron surface of an ideal quadrupole, ignoring the asymptotic tails.

where the constant k identifies a particular equipotential line. The magnetic field for the dipole is

$$B_y = -\frac{\partial A_z}{\partial x} = -c_1,$$

which is a constant.

The iron surfaces for higher order ideal multipoles can be found in a similar manner. The ideal quadrupole ($n = 2$) has

$$\begin{aligned} W &= c_2 z^2 &= c_2(x^2 - y^2 + 2\,i\,x\,y) \\ A_z &= c_2(x^2 - y^2) \\ \mu_0 V_m &= 2c_2 x\,y. \end{aligned}$$

The iron surface for the ideal quadrupole is the hyperbola

$$x\,y = \frac{h}{2c_2},$$

as shown in Figure 6.2. In polar coordinates, we can write the equation of the hyperbolic surface as

$$r^2 \sin 2\theta = a^2,$$

where a is the radius to the center of a pole and θ is measured from the positive x axis. There are four poles around the perimeter of the magnet that alternate in polarity. The surface hyperbola for a normal quadrupole has asymptotes along the x and y axes. The field components inside the aperture are

$$\begin{aligned} B_x &= -2c_2 y \\ B_y &= -2c_2 x. \end{aligned}$$

The vertical field on the midplane varies linearly across the aperture and is an example of a gradient field.

The symmetry properties of a magnet are determined by the symmetry of the poles. In an ideal normal $2N$-multipole, the poles are located at the azimuthal angles

$$\phi_k = (2k - 1)\frac{\pi}{2N}, \qquad k = 1, 2, \ldots, 2N \qquad (6.1)$$

The polarities of the poles alternate in direction. The spacing between the poles is π/N.

6.2 Approximate multipole configurations

It is not possible to build an ideal multipole magnet because the equipotential surfaces extend to infinity. Thus one is faced with approximating the ideal surface as well as possible to meet the field quality requirements for the magnet. Any approximation leads to the presence of additional allowed multipoles. The strength of the normal multipoles are proportional to $\cos \phi_k$. The conductor must be wound around the poles in such a way that the polarities of adjacent poles are in opposite directions. In order to get the poles to alternate in sign, we need

$$\cos \left(\phi_k + \frac{\pi}{N}\right) = -\cos \phi_k.$$

This is the same requirement that we saw in Section 4.6 for current distributions, so the allowed multipole components m are again given by

$$m = N (2n + 1), \qquad n = 0, 1, 2, \ldots$$

Halbach has described methods for determining the effects on the multipole coefficients of iron saturation and perturbations in the fabrication or construction of iron-dominated magnets.[3, 4] These methods involve determining the effect of the perturbation on the scalar potential associated with the pole surface. Among the effects he considers are azimuthal and radial displacements of the poles and modifications in the shape of the pole surface. For example, the addition of an iron shim with thickness profile $h(\phi)$ modifies the unperturbed scalar potential approximately by

$$\delta V_m \approx - h(\phi) H_r(\phi),$$

where $H_r(\phi)$ is the field on the unperturbed surface.

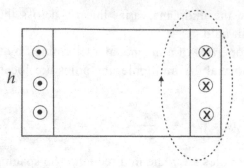

Figure 6.3 Window frame dipole.

6.3 Dipole configurations

Dipole magnets are commonly used to bend charged particle beams and for experiments requiring a uniform field.[5] The *window frame* dipole, shown in Figure 6.3, is a common configuration.[2, 6, 7] The coils approximate two parallel infinite current sheets, which we saw in Equation 4.33 produces a uniform vertical field. In the window-frame approximation, the field is very uniform across the aperture up to the vicinity of the coils. The field inside the coils falls off approximately linearly, reaching zero at the outer edge of the coils. If we look at the Ampère law around the dotted path indicated in Figure 6.3, we find that

$$NI = \int \vec{H} \cdot \vec{dl}$$

$$= \frac{B_0 h}{\mu_0} + \frac{B_0 L_{iron}}{\mu}, \tag{6.2}$$

where NI is the number of amp-turns in the coil, B_0 is the field on the midplane at the center of the aperture, h is the gap between the iron boundaries, and L_{iron} is the path length in the iron. Since $\mu \gg \mu_0$ and the typical path length in the iron is at most a few times greater than the gap, we have

$$\frac{h}{\mu_0} \gg \frac{L_{iron}}{\mu}.$$

The field produced by a window frame dipole is then

$$B_0 \simeq \frac{\mu_0 NI}{h}. \tag{6.3}$$

Note that the field strength is inversely proportional to the size of the gap.

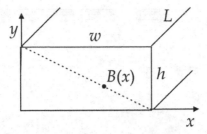

Figure 6.4 Cross-section through a dipole coil. The aperture of the magnet is located at negative x in this figure.

For finite permeability, the iron will undergo saturation at high field strengths, the second term in Equation 6.2 may no longer be negligible, and the field in the gap will be smaller than indicated by Equation 6.3. The field is lower at the center of the magnet compared to the field at the edges. This creates a small positive sextupole component in the field. Increasing the width of the pole beyond the useful aperture can improve the field quality.

Figure 6.4 shows a cross-section through one of the coils. The force on the conductor is

$$\vec{F} = \int \vec{J} \times \vec{B} \; dV.$$

The current density can be written as

$$J = \frac{NI}{wh} = \frac{B_0}{\mu_0 w}.$$

Assuming the field falls off linearly across the coil

$$B(x) = B_0 \frac{x}{w}$$

and using

$$dV = hL \; dx,$$

we can write [1]

$$F = \frac{B_0^2 \, hL}{\mu_0 \, w^2} \int_0^w x \, dx$$

$$= \frac{B_0^2 \, hL}{2 \, \mu_0}.$$

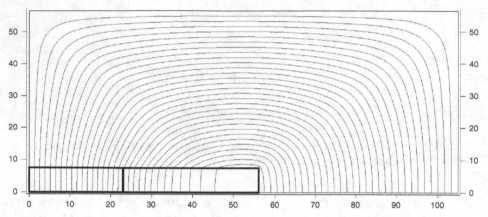

Figure 6.5 POISSON model of a window frame dipole. The dimensions are in centimeters.

Thus the force per unit length acting on the coil is

$$\frac{F}{L} = \frac{B_0^2 \, h}{2\mu_0}$$

and the transverse pressure is

$$P = \frac{B_0^2}{2\mu_0}.$$

Field calculations involving finite permeability iron have to be done using computer programs. Figure 6.5 shows a model[1] of a window frame dipole made with the program POISSON.[2] This figure shows one quarter of the cross-section of the magnet. The box in the lower left corner is an air region, which is the useful aperture in the magnet. The box to the right of the aperture is the conductor region. The remaining region is assumed here to be made of 1010 alloy steel. The contour lines show the direction of the magnetic field, which are vertical and fairly uniform in the aperture. POISSON breaks the iron into a large grid of points where the vector potential is computed. The relative permeability at each grid point is determined from a *B-H* curve. The one used here has a maximum μ_r value of 2,755. The program produces a self-consistent solution of Maxwell's equations. Table 6.1 summarizes the results for three values of the field B_0 at the center of the aperture. The second column shows the minimum value of the relative permeability

---

[1] This is a model of the 18D72 bending magnet that was built at Brookhaven National Laboratory.
[2] We will discuss the POISSON program in more detail in Chapter 11.

Table 6.1 *Summary of POISSON calculations for the window frame dipole*

B_0 [T]	Minimum μ_r	$x_{0.001}$ [cm]	F_3
1.56	65	21.2	$3.5\ 10^{-5}$
2.07	13	9.3	$4.0\ 10^{-4}$
2.59	4	3.0	$3.1\ 10^{-3}$

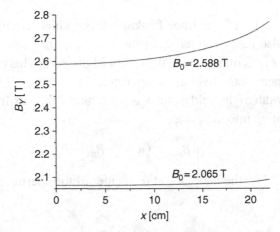

Figure 6.6 Magnetic field along the midplane aperture for the window frame dipole.

at any of the iron grid points. This quantity depends on the field strength in the iron and becomes smaller as the iron saturates at higher fields.

The third column shows the half-width of the good field region, defined here as the distance at which the field exceeds B_0 by more than 10^{-3} T. The last column shows the fractional contribution of the sextupole compared to the dipole contribution to the field. The strength of the field across the half-aperture is shown in Figure 6.6 for the two higher values of B_0. The field is smallest at the center of the aperture and grows as it approaches the conductor.

Another common dipole configuration is the *H*-dipole,[2, 6, 7] shown in Figure 6.7. The coils are recessed and hidden from direct view of the useful part of the magnet aperture. This makes the field less sensitive to errors in the coil location. The field is not as uniform as that in the window frame configuration. The error multipoles in a fixed, useful aperture decrease exponentially with increasing pole width. The iron near the edge of the pole is the first area that exhibits saturation. The good field region can also be extended by adding or subtracting material at the outer edges of the pole, rounding the corners, or by

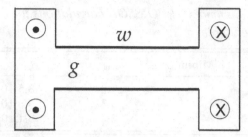

Figure 6.7 Cross-section of an *H*-dipole.

tapering the side edge of the iron. Leakage flux, which circulates around the conductors, can also cause saturation in the iron near the coils, so the magnetic field in the pole piece is maximum at the center of the pole. This creates a negative sextupole component in the field in the aperture.

The effective width of the field is $\approx w + g$. This causes the flux from the midplane to get squeezed going into the poles,

$$wB_{pole} \approx (w + g)B_0,$$

where B_0 is the field on the midplane at the center of the aperture. The field on the pole is then

$$B_{pole} \approx \left(1 + \frac{g}{w}\right)B_0.$$

The *H*-dipole has good field quality and mechanical stability. Simple racetrack-shaped coils can be used to excite the field in the iron.

The *C*-dipole, shown in Figure 6.8, is a configuration that allows good access to the magnet aperture from the side.[2, 6] The field in the iron can be excited with simple racetrack coils. However, the requirement for accessibility leads to a number of disadvantages. The necessary volume of the iron yoke is larger than for an *H*-dipole. There may be considerable leakage flux surrounding the conductors. The asymmetry in the yoke makes the mechanical stability worse. At high field levels, the attractive force between the poles can be quite large. Shims may be required at the edges of the poles to get acceptable field quality. There is a nonuniform magnetic field across the aperture, although this may be a desirable feature for applications that require a gradient field component. The field is smaller on the outside side of the gap than on the inside. The lack of left-right symmetry allows even harmonics to also be present in the field between the pole pieces. The fringe fields between the pole pieces extend outward by about a gap length on both sides of the pole pieces.

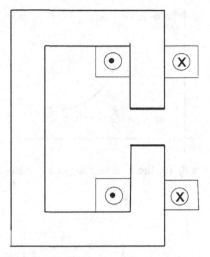

Figure 6.8 Cross-section of a *C*-dipole.

The *effective length* of a dipole, taking into account its end windings, is

$$L_{eff} = \frac{1}{B_0} \int_{-\infty}^{\infty} B(z)\, dz$$

$$\simeq L_{iron} + h,$$

(6.4)

where B_0 is the dipole strength in the center of the magnet. The quantity h is a length proportional to the aperture of the dipole, which takes into account the fringe field extending beyond the iron.

Conformal mapping techniques have been used to improve the modelling of the fringe field from dipole magnets.[8] Maps were used to transform the field from single and double-sided pole pieces with a uniform gap to the upper half of the complex plane.

6.4 Quadrupole configurations

Quadrupole magnets are frequently used for focusing charged particle beams.[5] We saw in Section 6.1 that an ideal quadrupole requires an infinitely long hyperbolic iron surface. A common method for terminating the iron boundary is to use symmetric cutoff angles θ_1, as shown for half of a symmetric pole in Figure 6.9.[6] The surface at the cutoff angles proceeds outwards along a radius. The equation of the hyperbolic surface relative to the centerline of the pole is

$$r^2 \cos 2\theta = a^2,$$

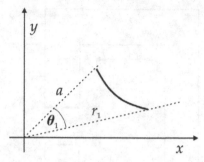

Figure 6.9 Cutoff angle θ_1 for the iron surface in a quadrupole magnet.

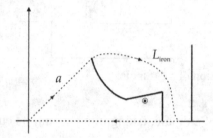

Figure 6.10 Loop through the quadrupole.

where a is the radius to the center of the pole. The cutoff angle θ_1 can be selected to be ~27° in order to eliminate the first allowed error multipole B_6. In this case, the radius to the cutoff point is given by

$$\frac{r_1}{a} \simeq 1.12.$$

The choice of the angle θ_1 also determines the amount of space available for the conductor. Excitation of the iron poles by the conductor can be determined by using the Ampère law around the path shown in Figure 6.10.

$$NI \simeq \int_0^a \frac{B(r)}{\mu_0}\, dr + \frac{B_0 L_{iron}}{\mu}.$$

The contribution from the path in the iron may be neglected since $\mu_r \gg 1$. The contribution for the path along the x axis vanishes because the field is perpendicular to the path. Thus we have,

$$NI \simeq \int_0^a \frac{gr}{\mu_0}\, dr$$

$$= \frac{ga^2}{2\mu_0},$$

where NI is the amp-turns around a pole and g is the quadrupole field gradient. Thus the gradient is given by

$$g = \frac{2\mu_0 NI}{a^2} \qquad (6.5)$$

and the pole tip field is

$$\begin{aligned}B_{pole} &= ga \\ &= \frac{2\mu_0 NI}{a}.\end{aligned} \qquad (6.6)$$

Saturation in the iron affects the area near the conductor first.

Figure 6.11 shows a POISSON model[3] of a quadrupole with hyperbolic pole pieces. The figure shows 1/8 of the symmetric cross-section. The 45° boundary splits one of the four poles in half. The region in the vicinity of the origin is the open aperture. The rectangular box on the side of the pole for $x \sim 14$ to 26 cm is the conductor, which wraps around the pole and returns on the opposite side of the symmetric half pole piece. The pole piece is part of the iron yoke that provides a flux path to the symmetric adjacent pole.

High-field dipoles and quadrupoles require pole piece materials with a large value for the saturation magnetic flux density. A number of soft magnetic materials with large B_{sat} are listed in Table 6.2. Also listed are the initial and peak values for the permeability and the coercivity. The resistivity of the material is important for considerations of eddy current losses in time-varying operations.

Quadrupoles have also been constructed by approximating the hyperbolic surface with a circular cylinder.[6] Consider a circle tangent to the hyperbola at the center of the pole, as shown in Figure 6.12. The circular surface is continued out to a cut-off angle θ_1 with respect to the center of the pole and then extends outward along a radius. Let a be the shortest distance from the center of the magnet to the pole and R be the radius of curvature of the circle, which is centered at C. Then

$$\begin{aligned}R &= R \sin \theta_1 + a \sin \theta_1 \\ &= \frac{a \sin \theta_1}{1 - \sin \theta_1}.\end{aligned}$$

The radius of the cutoff point is

$$r_1 = (R + a)\cos \theta_1.$$

---

[3] This model is an example file that is part of the POISSON code distribution.

Table 6.2 *Magnetic alloys with large B_{sat} [9]*

Alloy	Composition[1]	B_{sat} [T]	Initial μ_r	Max μ_r	H_c [Oe][2]	ρ_e [$\mu\Omega$-cm]
	35Co,1Cr	2.42	650	10,000	0.63	20
Supermendur	49Co,2 V	2.40	800	70,000	0.23	40
Vanadium permendur	49Co,2 V	2.35	800	6,000	2.20	40
Iron		2.14	150	5,000	1.00	10
Silicon steel	0.5Si	2.05	280	3,000	0.90	28
silicon steel	3Si	2.01	290	8,000	0.70	47
grain-oriented Si steel	3Si	2.01	1,400	50,000	0.09	50

[1] In percent, balance is Fe; [2] 1 Oe = 1 10^{-4} T / μ_0.

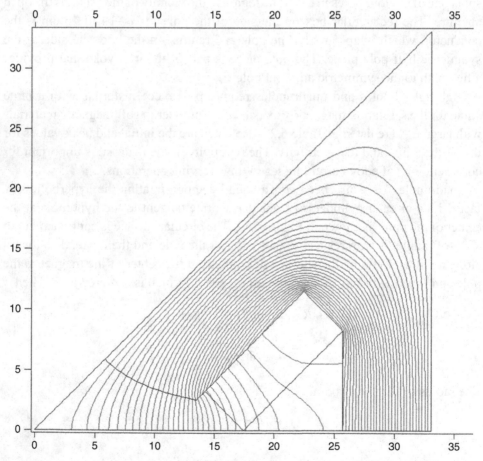

Figure 6.11 POISSON model of a quadrupole with hyperbolic pole pieces. Dimensions are in centimeters.

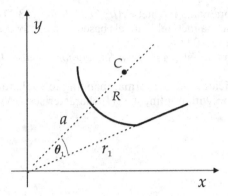

Figure 6.12 Quadrupole magnet with circular iron surface.

One solution to these equations, which makes the first allowed multipole harmonic $B_6 = 0$, is

$$\theta_1 = 31.5°$$

$$\frac{r_1}{a} = 1.785$$

$$\frac{R}{a} = 1.094.$$

Conformal mapping techniques have been used to simplify the design of quadrupoles and higher order multipole magnets.[10] The desired higher-order multipole is mapped to a dipole geometry, where it is easier to understand what effects proposed modifications make to the field in the useful aperture. It is also possible to numerically determine the field quality in the higher multipole aperture more accurately by computing the multipole coefficients in the transformed geometry.

References

[1] J. Tanabe, *Iron-Dominated Electromagnets*, World Scientific, 2005.
[2] T. Zickler, Basic design and engineering of normal-conducting, iron-dominated electromagnets, in *Proc. CERN Accelerator School on Magnets*, Bruges, Belgium, CERN-2010-004, June 2009, p. 65.
[3] K. Halbach, First order perturbation effects in iron dominated two dimensional symmetrical magnets, *Nuc. Instr. Meth.* 74:147, 1969.
[4] K. Halbach, Fields and first order perturbation effects in two dimensional conductor dominated magnets, *Nuc. Instr. Meth.* 78:185, 1970.
[5] A. Banford, *The Transport of Charged Particle Beams*, Spon Ltd., 1966, p. 81 84, 100–103.
[6] G. Parzen, Magnetic Fields for Transporting Charged Beams, Brookhaven National Laboratory Report 50536, 1976.

[7] G. Fischer, Iron-dominated magnets, *AIP Conf. Proc.* 153:1147–1151, 1987.

[8] P. Walstrom, Dipole magnet field models based on a conformal map, *Phys. Rev.-Spec. Top. Accel. Beams* 15:102401, 2012.

[9] Magnetically soft materials, *American Society for Metals (ASM) Handbook*, vol. 2, 1990, p. 761–781.

[10] K. Halbach, Application of conformal mapping to evaluation and design of magnets containing iron with nonlinear $B(H)$ characteristics, *Nuc. Instr. Meth.* 64:278, 1968.

7

Axial field configurations

In this chapter, we consider field configurations that have an axial field component. In straight channels, these fields are azimuthally symmetric around the system axis and only have axial and radial components. The basic example of this type of configuration is the closed circular current loop. Combinations of current loops can be used to produce desired axial field profiles. The current loop can also be extended axially to generate an ideal sheet solenoid. We conclude the chapter with a discussion of bent solenoids. When the bent channel forms a closed ring, we obtain the toroid configuration.

7.1 Circular current loop

We recall from Equation 1.18 that the on-axis field of a circular current loop with radius a is

$$B_z = \frac{\mu_0 I a^2}{2\{a^2 + z_o^2\}^{3/2}}, \tag{7.1}$$

where I is the current in the loop and z_o is the distance of the observation point along the z axis from the plane of the loop. We now consider the determination of the vector potential in the case when the observation point P is not restricted to lie along the z axis, as shown in Figure 7.1. We define a coordinate system where the x axis lies directly below the observation point P. By symmetry, the vector potential only has a ϕ component and cannot depend on the azimuthal coordinate ϕ. The vector potential is given by

$$A_\phi(\rho, z) = \frac{\mu_0 I}{4\pi} \oint \frac{ds}{R}.$$

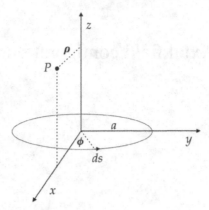

Figure 7.1 Current loop geometry.

An arbitrary element of current has the Cartesian coordinates $(a \cos \phi, a \sin \phi, 0)$, so

$$R = \left\{ (\rho - a \cos \phi)^2 + (a \sin \phi)^2 + z^2 \right\}^{1/2}$$
$$= \left\{ \rho^2 + a^2 - 2a\rho \cos \phi + z^2 \right\}^{1/2}.$$

The contribution from a current element at ϕ makes the same contribution to the vector potential as the element at $-\phi$. In addition, the contribution of each of these elements to the vector potential at P is proportional to $\cos \phi$. Therefore we can write

$$A_\phi(\rho, z) = \frac{\mu_0 I}{2\pi} \int_0^\pi \frac{\cos \phi}{\{\rho^2 + a^2 - 2a\rho \cos \phi + z^2\}^{1/2}} a \, d\phi.$$

Making the substitutions

$$\phi = \pi + 2\theta$$
$$\cos \phi = -1 + 2\sin^2 \theta,$$

we can write the vector potential as

$$A_\phi(\rho, z) = \frac{\mu_0 I a}{2\pi} \int_{-\pi/2}^0 \frac{2\sin^2\theta - 1}{\{\rho^2 + a^2 + z^2 - 2a\rho(2\sin^2\theta - 1)\}^{1/2}} 2 \, d\theta.$$

The integral is symmetric in θ, so we can translate the limits of integration. After rearranging the terms in the denominator, we get

$$A_\phi(\rho, z) = \frac{\mu_0 I a}{\pi} \int_0^{\pi/2} \frac{2\sin^2\theta - 1}{\{(a + \rho)^2 + z^2 - 4a\rho \sin^2\theta\}^{1/2}} \, d\theta.$$

Define

$$k^2 = \frac{4a\rho}{(a+\rho)^2 + z^2}.$$ (7.2)

Then we have

$$A_\phi(\rho, z) = \frac{\mu_0 I a}{\pi} \frac{k}{\sqrt{4a\rho}} \int_0^{\pi/2} \frac{2\sin^2\theta - 1}{\{1 - k^2\sin^2\theta\}^{1/2}} \, d\theta$$

$$= \frac{\mu_0 I a}{\pi} \frac{k}{\sqrt{4a\rho}} [2\,\mathbb{I}_1 - \mathbb{I}_2],$$ (7.3)

where[1]

$$\mathbb{I}_1 = \int_0^{\pi/2} \frac{\sin^2\theta}{\{1 - k^2\sin^2\theta\}^{1/2}} \, d\theta$$

$$= \frac{K(k) - E(k)}{k^2}$$

and[2]

$$\mathbb{I}_2 = \int_0^{\pi/2} \frac{1}{\{1 - k^2\sin^2\theta\}^{1/2}} \, d\theta$$

$$= K(k).$$

The function $K(k)$ is the complete elliptic integral of the first kind and $E(k)$ is the complete elliptic integral of the second kind.[3] Substituting back into Equation 7.3, we find

$$A_\phi(\rho, z) = \frac{\mu_0 I a}{\pi} \frac{k}{\sqrt{4a\rho}} \left[\left(\frac{2}{k^2} - 1\right) K(k) - \frac{2}{k^2} E(k) \right],$$

which can be written in the form [1, 2]

$$A_\phi(\rho, z) = \frac{\mu_0 I}{\pi k} \sqrt{\frac{a}{\rho}} \left[\left(1 - \frac{k^2}{2}\right) K(k) - E(k) \right].$$ (7.4)

---

[1] GR 8.112.5.   [2] GR 8.112.1.   [3] Properties of complete elliptic integrals are discussed in Appendix F.

The components of the magnetic field in cylindrical coordinates are

$$B_\rho = -\frac{\partial A_\phi}{\partial z}$$

$$B_\phi = 0$$

$$B_z = \frac{1}{\rho}\frac{\partial}{\partial \rho}(\rho A_\phi).$$

In order to evaluate these field components, we need the derivatives of the parameter k defined in Equation 7.2. We have [1]

$$\frac{\partial k}{\partial z} = -\frac{zk^3}{4a\rho}$$

$$\frac{\partial k}{\partial \rho} = \frac{k}{2\rho} - \frac{k^3}{4\rho} - \frac{k^3}{4a}. \tag{7.5}$$

We also need the derivatives[4] of the complete elliptic integrals $K(k)$ and $E(k)$ with respect to their parameter k.

$$\frac{\partial K}{\partial k} = \frac{E}{k\,(1-k^2)} - \frac{K}{k}$$

$$\frac{\partial E}{\partial k} = \frac{E}{k} - \frac{K}{k}. \tag{7.6}$$

Evaluating the derivatives together with a lot of algebra,[5] we find that [1, 2]

$$B_\rho = \frac{\mu_0 I}{2\pi} \frac{z}{\rho\{(a+\rho)^2 + z^2\}^{1/2}} \left[-K(k) + \frac{a^2 + \rho^2 + z^2}{(a-\rho)^2 + z^2} E(k) \right] \tag{7.7}$$

and

$$B_z = \frac{\mu_0 I}{2\pi} \frac{1}{\{(a+\rho)^2 + z^2\}^{1/2}} \left[K(k) + \frac{a^2 - \rho^2 - z^2}{(a-\rho)^2 + z^2} E(k) \right]. \tag{7.8}$$

In the limit $\rho \to 0$, $k^2 = 0$ and the elliptic integrals in Equation 7.8 equal $\pi/2$. Then it is straightforward to show that B_z approaches Equation 7.1 for the axial field on the axis. Using l'Hopital's rule and the series expansions

---

[4] GR 8.123.2,4.   [5] A computer algebra program is really useful here!

$$E(k) \simeq \frac{\pi}{2} - \frac{\pi}{8} k^2 + \cdots$$

$$K(k) \simeq \frac{\pi}{2} + \frac{\pi}{8} k^2 + \cdots,$$

it is also possible to show that Equation 7.7 for B_ρ approaches 0 on the axis, as it should.

In the preceding derivation, we have gone through a standard approach of calculating the vector potential and taking its derivatives to find the field components. We have done this to illustrate several useful mathematical properties involving the use of elliptic integrals. We should note, however, that it is possible in this case to solve the Biot-Savart equation for the fields directly since the required integrals are known.[3]

Besides the solution given here in terms of $K(k)$ and $E(k)$ and cylindrical coordinates, the problem of the circular current loop has been solved using a number of alternative methods. The vector potential for the circular loop can be written in terms of Bessel functions as [4]

$$A_\phi(\rho, z) = \frac{\mu_0 I \, a}{2} \int_0^\infty J_1(ka) \, J_1(k\rho) \, e^{-k|z|} \, dk. \tag{7.9}$$

For some applications, it is more convenient to solve for the vector potential of the current loop in spherical coordinates. Spherical solutions for the vector potential and field have also been given in terms of elliptic integrals.[5, 6] However in spherical coordinates, it is sometimes more natural to expand the solutions in Legendre functions. The vector potential for the current loop for $r < a$ is given in this case as [1]

$$A_\phi(r, \theta) = \frac{\mu_0 I}{2} \sum_{n=1}^\infty \frac{\sin \alpha}{n(n+1)} \left(\frac{r}{a}\right)^n P_n^1(\cos \alpha) \, P_n^1(\cos \theta), \tag{7.10}$$

where P_n^1 is an associated Legendre function and α is the polar angle of the loop. For $r > a$, the radial factor in this equation must be replaced with

$$\left(\frac{a}{r}\right)^{n+1}.$$

This type of expansion makes it easier to show that the field of the current loop approaches that for a magnetic dipole in the limit when $r \gg a$. It is also possible to solve for the field components directly from Maxwell's equations.[7]

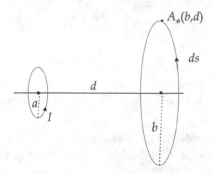

Figure 7.2 Mutual inductance of two current loops.

Example 7.1: mutual inductance of two coaxial current loops
Consider two coaxial current loops separated by a distance d, as shown in Figure 7.2. The mutual inductance between the two loops is the flux intercepted by loop 2 for a given current in loop 1. Thus we have

$$M = \frac{\Phi_2}{I_1} = \frac{1}{I_1} \int A_{\phi 1}(b, d) \, ds_2.$$

The vector potential for loop 1 at points along loop 2 can be found using Equation 7.4

$$A_{\phi}(b, d) = \frac{\mu_0 I}{\pi k} \sqrt{\frac{a}{b}} \left[\left(1 - \frac{k^2}{2} \right) K(k) - E(k) \right],$$

where k^2 follows from Equation 7.2.

$$k^2 = \frac{4ab}{(a + b)^2 + d^2}.$$

Since the value of A_{ϕ} is constant for all the points on loop 2, the mutual inductance is [8]

$$M = \mu_0 \sqrt{ab} \left[\left(\frac{2}{k} - k \right) K(k) - \frac{2}{k} E(k) \right]. \tag{7.11}$$

7.2 Radial expansion of the on-axis magnetic field

Consider a longitudinal distribution of azimuthally symmetric current sources. It is useful in some cases to express the off-axis values of the magnetic field as a function of the magnetic field along the system axis of the current distribution. This could be used, for example, to synthesize some desirable field distribution or for rapid optimization of current source parameters. We begin by assuming we have a scalar potential

$$\Omega = \mu_0 V_m$$

that does not depend on the coordinate ϕ and can be written as the power series

$$\Omega(\rho, z) = \sum_{n=0}^{\infty} c_n(z)\, \rho^n.$$

We demand that Ω satisfy Laplace's equation in the region from the axis up to the location of the closest coil

$$\nabla^2 \Omega = 0,$$

where the Laplacian operator is given in cylindrical coordinates by

$$\nabla^2 = \frac{1}{\rho}\, \partial_\rho (\rho\, \partial_\rho) + \partial_z^2.$$

Substituting the series for Ω into Laplace's equation and bringing the operator inside the summation sign, we get

$$\sum_n \left[c_n\, n^2 \rho^{n-2} + \rho^n\, \frac{\partial^2 c_n}{\partial z^2} \right] = 0.$$

In order to satisfy this relation, we need to get cancellations between the c_n terms of order n and second derivative terms two orders higher than n. Therefore let us demand that

$$c_{n+2}(n+2)^2 \rho^n = -\rho^n\, \frac{\partial^2 c_n}{\partial z^2}.$$

Making the substitution $n \rightarrow n-2$, we can write the coefficient in terms of the recursion relation

$$c_n = -\frac{1}{n^2}\, \frac{\partial^2 c_{n-2}}{\partial z^2}, \qquad n \geq 2. \tag{7.12}$$

We know that the radial component of the magnetic field has to vanish on the axis of the system. Since

$$B_\rho = -\frac{\partial \Omega}{\partial \rho},$$

we find that

$$-B_\rho = c_1 + 2c_2\,\rho + 3c_3\,\rho^2 + \cdots$$

Therefore, we must have $c_1 = 0$, and since Equation 7.12 relates c_1 to all the higher odd terms, the series expansion for Ω can only contain even terms. Thus Ω has the form

$$\Omega(\rho, z) = \sum_{n=0}^{\infty} c_{2n}(z)\,\rho^{2n},$$

where

$$c_{2n}(z) = \frac{(-1)^n \dfrac{\partial^{2n} c_0}{\partial z^{2n}}}{2^{2n}(n!)^2}. \tag{7.13}$$

The numerical factors in the coefficient can be checked by comparing the values from Equation 7.13 with the values from recursively using Equation 7.12. Define the magnetic field on the system axis as

$$B_0(z) = B_z(0, z) = -\left.\frac{\partial \Omega}{\partial z}\right|_{\rho=0} = -\frac{\partial c_0}{\partial z}.$$

Then the off-axis axial field component is [9]

$$B_z(\rho, z) = \sum_{n=0}^{\infty} \frac{(-1)^n}{2^{2n}(n!)^2} \frac{\partial^{2n} B_0}{\partial z^{2n}}\,\rho^{2n} \tag{7.14}$$

and the off-axis radial field component is

$$B_\rho(\rho, z) = \sum_{n=0}^{\infty} \frac{(-1)^{n+1}}{2^{2n+1} n!(n+1)!} \frac{\partial^{2n+1} B_0}{\partial z^{2n+1}}\,\rho^{2n+1}. \tag{7.15}$$

In cases involving loops and solenoids, where the on-axis fields are known analytically, it is possible using this method to achieve high accuracy in computing the field out to radial distances ~70% of the coil radius.[9]

7.3 Zonal harmonic expansions

The solution of Laplace's equation in spherical coordinates for azimuthally symmetric current distributions can be expressed as a series of *zonal harmonic*

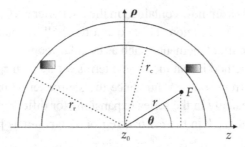

Figure 7.3 Geometry for zonal harmonic calculations.

functions.[10, 11] Computations of magnetic fields using these expansions can be faster than calculations using elliptic integrals. In addition, the harmonic expansion allows easier optimization, where, for example, we can get better field quality by eliminating leading error terms in the series.

Consider the spherical coordinate system shown in Figure 7.3. The z axis is the polar axis of symmetry. We choose a *source point* z_0 along the z axis as the origin of the coordinate system. We define a *central region* extending from the origin to a radius r_c that is the shortest distance to the edge of any current element. We define the *remote region* to extend from the radius r_r, which is the longest distance from the origin to any part of a current element, to infinity. The zonal harmonic expansion can be written as convergent series for $r < r_c$ and for $r > r_r$. The expansion for a given source point is divergent in the region $r_c < r < r_r$. However, it is possible to extend the region of validity by moving the source point. The conductors are azimuthally symmetric around the z axis. The field point F is defined to have the spherical coordinates r and θ.

The magnetic scalar potential $V = V_m$ is a solution of Laplace's equation. Let us define $u = \cos \theta$. The zonal harmonic solution in the central region is

$$V = \sum_{n=0}^{\infty} c_n \, r^n P_n(u),$$

where $P_n(u)$ is the Legendre polynomial[6] of order n. We choose to write the coefficients c_n in the following manner in order to simplify the expressions for the magnetic field.

$$c_0 = V(z_0)$$
$$c_n = -\frac{1}{\mu_0 n \, r_c^{n-1}} B_{n-1}^c, \quad n > 0.$$

---

[6] Some important properties of Legendre functions are discussed in Appendix D.

The unknown quantities are now contained in the coefficients B_n^c, which are called the *source terms* for the central region. These quantities will be related later to the fields produced by various current elements, such as circular loops. An important feature of the zonal harmonic method is that the source terms for a given z_o only depend on the coordinates of the current source. Thus once the source terms have been calculated, they can be used repeatedly in the series expansions for different field points.

With these definitions, V in the *central* region is written as [10]

$$V = V(z_o) - \frac{r_c}{\mu_0} \sum_{n=1}^{\infty} \frac{B_{n-1}^c}{n} \left(\frac{r}{r_c}\right)^n P_n(u).$$

In order to compute the cylindrical components of the magnetic field, we first make use of the following relations from Figure 7.3.

$$r = \sqrt{\rho^2 + (z - z_o)^2}$$
$$\partial_z r = u$$
$$\partial_\rho r = \sin \theta$$

and

$$u = \frac{z - z_o}{r}$$
$$\partial_z u = \frac{1 - u^2}{r}$$
$$\partial_\rho u = -\frac{u}{r} \sin \theta.$$

The axial component of the field is

$$B_z = r_c \sum_{n=1}^{\infty} \frac{B_{n-1}^c}{n \, r_c^n} \left[r^n \, P_n'(u) \left(\frac{1 - u^2}{r}\right) + P_n(u) n \, r^{n-1} \, u \right],$$

where P'_n is the derivative of P_n with respect to its argument u. The quantity in square brackets is

$$[\,] = r^{n-1} \{ P_n'(u) (1 - u^2) + nu P_n(u) \}.$$

We can use the recursion relation for Legendre polynomials[7]

$$(1 - u^2) P_n' = n P_{n-1} - un P_n \tag{7.16}$$

---

[7] GR 8.914.2.

to write

$$[\,] = r^{n-1} n P_{n-1}(u).$$

Substituting back into the equation for B_z, we find

$$B_z = \sum_{n=1}^{\infty} B_{n-1}^c \left(\frac{r}{r_c}\right)^{n-1} P_{n-1}(u).$$

Changing the index to $m = n - 1$, the axial field in the central region can be written as

$$B_z = \sum_{m=0}^{\infty} B_m^c \left(\frac{r}{r_c}\right)^m P_m(u). \tag{7.17}$$

Following a similar procedure, the transverse field component in the central region is

$$B_\rho = r_c \sum_{n=1}^{\infty} \frac{B_{n-1}^c}{n \; r_c^n} [r^{n-1} \sin \theta \; (-u P_n' + n P_n)].$$

Using the recursion relation [12]

$$n P_n = u P_n' - P_{n-1}'$$

and shifting the series index again, we find the transverse field in the central region is

$$B_\rho = -\sin \theta \sum_{m=0}^{\infty} \frac{B_m^c}{m + 1} \left(\frac{r}{r_c}\right)^m P_m'(u). \tag{7.18}$$

In the *remote* region, we write the scalar potential in the form

$$V = V_0(\pm\infty) + \frac{r_r}{\mu_0} \sum_{n=1}^{\infty} \frac{B_{n+1}^r}{n + 1} \left(\frac{r_r}{r}\right)^{n+1} P_n(u).$$

The axial field is

$$B_z = -\sum_{n=1}^{\infty} \frac{B_{n+1}^r}{n + 1} \left(\frac{r_r}{r}\right)^{n+2} \left[P_n'(1 - u^2) - P_n u \; (n + 1)\right].$$

Using the recursion relation Equation 7.16 and shifting the series index, we find the axial field component in the remote region is

$$B_z = \sum_{m=2}^{\infty} B_m^r \left(\frac{r_r}{r}\right)^{m+1} P_m(u). \tag{7.19}$$

The transverse field component in the remote region is

$$B_\rho = -\sum_{n=1}^{\infty} \frac{B_{n+1}^r}{n+1} \left(\frac{r_r}{r}\right)^{n+2} \sin\theta \left[-uP_n' - (n+1)P_n\right].$$

Using the recursion relation [13]

$$(n+1)P_n = P_{n+1}' - uP_n'$$

and shifting the series index, we find that the transverse field component in the remote region is

$$B_\rho = \sin\theta \sum_{m=2}^{\infty} \frac{B_m^r}{m} \left(\frac{r_r}{r}\right)^{m+1} P_m'(u). \tag{7.20}$$

Now that we have determined the series representations of the field components in the central and remote regions, we turn to the calculation of the *source* terms. Consider a field point on the z axis in the central region with $z > z_o$. In this case, $\theta = 0$, $\rho = 0$, $u = 1$, and $r = z - z_o$. From Equation 7.17, we have

$$B_0(z) = B_z(0, z) = \sum_{n=0}^{\infty} \frac{B_n^c}{r_c^n} (z - z_o)^n.$$

When $n = 0$ and $z = z_o$, we see that the coefficient

$$B_0^c = B_z(0, z_o)$$

is the axial magnetic field at the source point. The Taylor series around the point z_0 is

$$B_0(z) = \sum_{n=0}^{\infty} \frac{1}{n!} \frac{d^n B_0}{dz^n}\bigg|_{z_o} (z - z_o)^n. \tag{7.21}$$

Comparing the two series for B_0, we find that [11]

$$B_n^c = \frac{r_c^n}{n!} \frac{d^n B_0}{dz^n}(z_o). \tag{7.22}$$

Suppose that the current source is a circular loop, as shown in Figure 7.4. For a current loop, we have $r_c = r_r = r_L$. From Equation 7.1, the axial field at the field point F is

$$B_0(z) = \frac{\mu_0 I \rho_L^2}{2d^3}. \tag{7.23}$$

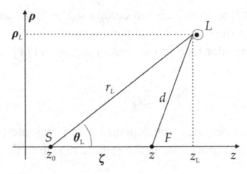

Figure 7.4 Source terms for a circular current loop.

Define $\zeta = z - z_o$. From the law of cosines, we have

$$d = \sqrt{r_L^2 + \zeta^2 - 2\,r_L\,\zeta\,u_L}.$$

We make use of the following series expansion for Legendre polynomials[8]

$$\frac{1}{\sqrt{1 + h^2 - 2hu}} = \sum_{n=0}^{\infty} h^n\,P_n(u),$$

where $h < 1$. Differentiating both sides of this equation with respect to u, we find

$$\frac{1}{\{1 + h^2 - 2hu\}^{3/2}} = \sum_{n=0}^{\infty} h^n\,P'_{n+1}(u). \tag{7.24}$$

In the central region, let

$$h = \frac{\zeta}{r_L} < 1,$$

so that

$$d = r_L\sqrt{1 + h^2 - 2\,hu_L}.$$

Using Equation 7.24, we find that

$$\frac{1}{d^3} = \frac{1}{r_L^3}\sum_{n=0}^{\infty} h^n\,P'_{n+1}(u_L).$$

[8] GR 8.921.

Substituting this into Equation 7.23 and comparing with the general series expansion Equation 7.17 for the case of a field point on the axis, we can conclude that the source term for the circular loop in the central region is [11]

$$B_n^c = \frac{\mu_0 I \rho_L^2}{2 r_L^3} P_{n+1}'(u_L).$$

(7.25)

Note that this expression does not depend on any parameters of a field point. We follow an analogous procedure in the remote region to find that

$$B_n^r = \frac{\mu_0 I \rho_L^2}{2 r_L^3} P_{n-1}'(u_L).$$

(7.26)

7.4 Multiple coil configurations

Combinations of current loops are often used to create regions of space with some desired magnetic properties. We can generalize the current source as a "small" coil with N turns, provided the size of the coil is small compared with the separation between the coils. In this case, we replace the loop current I in the field equations with the product NI. The classic example of a multiple coil configuration is the Helmholtz pair, where two coils are used to create a region of approximately uniform field.

Consider the arrangement of two coaxial current loops shown in Figure 7.5. The two loops are perpendicular to the z axis and have the same radius a and the same current I. The axial field of each loop is given by Equation 7.1. In a coordinate system with the origin midway between the coils, the axial field of the two loops can be written

$$B_z(0, z) = \frac{\mu_0 I \, a^2}{2} F(z),$$

(7.27)

where

$$F(z) = \frac{1}{\{a^2 + (b - z)^2\}^{3/2}} - \frac{1}{\{a^2 + (b + z)^2\}^{3/2}}.$$

(7.28)

In the Helmholtz configuration, the spacing $2b$ between the coils equals the radius a of the coils. The field of a Helmholtz pair at the origin is

$$B_z(0, 0) = \frac{\mu_0 I}{a} \frac{8}{5\sqrt{5}}$$

$$\simeq 0.7155 \, \frac{\mu_0 I}{a}.$$

(7.29)

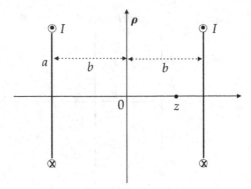

Figure 7.5 A two-coil configuration.

The axial field between the coils falls off slowly with z. At the center of the current loops the field is

$$B_z\left(0,\frac{a}{2}\right) = \frac{\mu_0 I}{a}\frac{1}{2}\left[1 + \frac{1}{\sqrt{8}}\right]$$

$$\simeq 0.6768\,\frac{\mu_0 I}{a}\,.$$

In the vicinity of the origin, the axial field can be expanded in the Taylor series, Equation 7.21. The field uniformity is determined by the leading-order terms in this expansion. Because of the symmetry of the coil arrangement, all the odd power terms in the series have to vanish. The virtue of the Helmholtz configuration is that the second derivative term in the expansion also vanishes. Thus the leading correction in the Taylor series is the fourth order term, which is proportional to

$$\frac{\partial^4 B_z}{\partial z^4} \simeq -19.8\,\frac{\mu_0 I}{a^5}\,.$$

Thus in the vicinity of the origin, the axial field is [14]

$$B(0,z) \simeq \frac{\mu_0 I}{a}\left[0.7155 - 0.825\left(\frac{z}{a}\right)^4 + \cdots\right]. \tag{7.30}$$

The field at any point off the axis can be found by using the elliptic integral solutions for the current loop given in Equations 7.7 and 7.8.[15] In the plane midway between the coils, the field only has an axial component because of symmetry. Defining the scaled radius $u = \rho/a$, the elliptic integral parameter is

$$k^2 = \frac{16u}{4u^2 + 8u + 5}$$

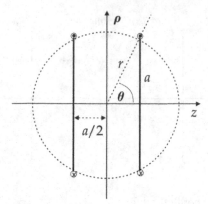

Figure 7.6 Helmholtz coil configuration.

and the axial field is

$$B_z(u,0) = \frac{2\mu_0 I}{\pi a} \frac{1}{\sqrt{4u^2 + 8u + 5}} \left[K(k) + \frac{3 - 4u^2}{4u^2 - 8u + 5} E(k) \right].$$

By numerically evaluating this expression, the field is found to be quite uniform in the vicinity of the axis.[16] It falls off to 99.93% of the on-axis value at a radius of 0.2a.

The Helmholtz pair arrangement has the geometric property that the two coils lie on the surface of a sphere, as illustrated in Figure 7.6. For this configuration, we have

$$\tan \theta = 2$$
$$\sin \theta = \frac{a}{r},$$

so the radius of the sphere is

$$r = \frac{a}{\sin(\tan^{-1} 2)} \simeq 1.118 \, a.$$

The Helmholtz pair also has interesting asymptotic behavior.[17] Expanding the on-axis field in powers of a/z, the field at large distance is given by

$$B_z(0,z) \simeq \frac{\mu_0 I a^2}{z^3} + \frac{3\mu_0 I a^2}{2z^5} (4b^2 - a^2) + \cdots$$

The leading term is the magnetic dipole term. However, the next term in the series vanishes under the Helmholtz condition $a = 2b$.

An *inverse Helmholtz* pair has the currents in the two coils flowing in opposite directions. In this case, the field at the origin vanishes, and the leading multipole term is the field gradient. The optimum gradient for fixed radius a is [18]

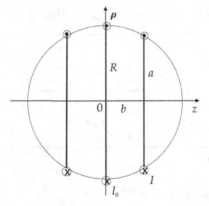

Figure 7.7 Maxwell tricoil configuration.

$$\frac{dB_z}{dz} = \frac{48}{25\sqrt{5}} \frac{\mu_0 I}{a^2}$$

$$\simeq 0.8587 \frac{\mu_0 I}{a^2}.$$

If practical constraints demand it, other gradient optimizations are possible for fixed spacing b or for fixed radius $r^2 = a^2 + b^2$.[19]

The classic design using three coils is the *Maxwell tricoil*, shown in Figure 7.7. The tricoil has a pair of identical coils and a third coil with larger radius in the symmetry plane between the coil pair. The three coils all lie on the surface of a sphere. This design improves on the field quality from the Helmholtz pair by also making the fourth-order term in the Taylor series vanish. Thus the first correction term is sixth-order. Maxwell's solution is

$$a = \sqrt{\frac{4}{7}} R$$

$$b = \sqrt{\frac{3}{7}} R$$

$$I = \frac{49}{64} I_0 .$$

The magnetic field at the origin is

$$B_z(0,0) = 60 \frac{\mu_0 I}{R}.$$

An improved three coil design with three circular coils of the same radius has a larger uniform field region than Maxwell's design.[15] Another sixth-order

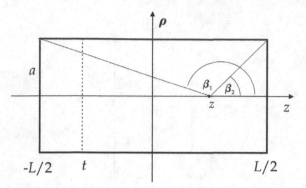

Figure 7.8 Sheet model of a solenoid.

design used three square coils.[20] An exhaustive study of multi-coil systems using the methods of zonal harmonics has identified many uniform and gradient field configurations with higher-order error corrections.[21]

7.5 Sheet model for the solenoid

We consider here the field from a solenoid in the approximation that the radial thickness of the solenoid coil can be neglected. We assume that the sheet conductor is composed of circular current loops that are extended in the axial direction for the length L of the solenoid, as shown in Figure 7.8. We compute the on-axis field at the location z of a finite length solenoid by integrating the field of a current loop

$$B_z(0, z) = \int_{-L/2}^{L/2} \frac{\mu_0 I a^2}{2\left\{a^2 + (z - t)^2\right\}^{3/2}} \, n \, dt \, ,$$

where n is the number of turns per unit length. Performing the integration[9] gives

$$B_z(0, z) = \frac{\mu_0 n I}{2} \left[\frac{z + L/2}{\left\{a^2 + (z + L/2)^2\right\}^{1/2}} - \frac{z - L/2}{\left\{a^2 + (z - L/2)^2\right\}^{1/2}} \right]. \qquad (7.31)$$

This can be written in the form

$$B_z(0, z) = \frac{\mu_0 n I}{2} \left(\cos \beta_2 - \cos \beta_1 \right), \qquad (7.32)$$

where β_i are the angles subtended at location z on the axis of the solenoid to the outer edges of the two ends of the current sheet. The field in the center of the solenoid is

[9] GR 2.264.5.

$$B_z(0,0) = \mu_0 n I \frac{L}{\sqrt{L^2 + 4a^2}}. \tag{7.33}$$

In the limit of an infinitely long current sheet, this expression reduces to the field of an ideal solenoid, Equation 1.28.

$$B_z = \mu_0 n I.$$

There is a close connection between the derivatives of the on-axis solenoid field and the on-axis fields of the current loops at the ends of the solenoid.[9, 10] In the coordinate system in Figure 7.8,

$$\frac{dB_z^{Solenoid}(0,z)}{dz} = n \left[B_z^{Loop}\left(0, z + \frac{L}{2}\right) - B_z^{Loop}\left(0, z - \frac{L}{2}\right) \right] \tag{7.34}$$

The off-axis expansion method discussed in Section 7.2 can be used in conjunction with Equation 7.34 to find the field of a sheet solenoid.[9]

We turn next to calculating the field of a solenoid at any point, including points off the symmetry axis. We will perform a direct calculation of the field using the Biot-Savart equation

$$\overrightarrow{dB} = \frac{\mu_0 I}{4\pi} \frac{\overrightarrow{dl} \times \overrightarrow{R}}{R^3}.$$

Consider the solenoid geometry shown in Figure 7.9. Because of the azimuthal symmetry of the current, the field is also azimuthally symmetric. Thus for

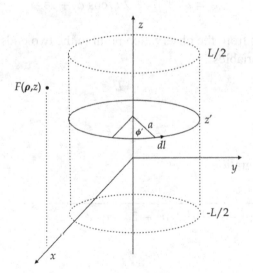

Figure 7.9 Geometry of the sheet solenoid.

mathematical simplicity, we are free to choose the field point F to lie directly above the x axis. The distance from the source current element to the field point is

$$\vec{R} = (\rho - a \cos \phi')\,\hat{x} - a \sin \phi'\,\hat{y} + (z - z')\,\hat{z}, \qquad (7.35)$$

while the current element is

$$\vec{dl} = -a \sin \phi'\, d\phi'\,\hat{x} + a \cos \phi'\, d\phi'\,\hat{y}. \qquad (7.36)$$

We first compute the axial component of the field. Taking the z component of the cross-product in the Biot-Savart equation, we find that the ϕ' dependence only involves terms in $\cos \phi'$. Thus symmetric current elements with respect to the x axis make identical contributions to the integral. We have

$$B_z = \frac{\mu_0 I' a}{2\pi} \int_{-L/2}^{L/2} \int_0^\pi \frac{a - \rho \cos \phi'}{\{a^2 + \rho^2 - 2a\rho \cos \phi' + (z - z')^2\}^{3/2}}\, dz'\, d\phi'$$

$$= \frac{\mu_0 I' a}{2\pi} \int_0^\pi (a - \rho \cos \phi')\, \mathbb{I}_1\, d\phi',$$

where I' is the sheet current density,

$$\mathbb{I}_1 = \int_{-L/2}^{L/2} \frac{dz'}{\{e - 2\, z\, z' + z'^2\}^{3/2}}$$

and we define

$$e = a^2 + \rho^2 - 2a\rho \cos \phi' + z^2. \qquad (7.37)$$

Define the distances from the observation point to the two ends of the solenoids in terms of the new variables

$$z_1 = -\frac{L}{2} - z$$

$$z_2 = \frac{L}{2} - z.$$

After doing the integration, we get[10]

$$\mathbb{I}_1 = \frac{z_2}{(e - z^2)\{a^2 + \rho^2 - 2a\rho \cos \phi' + z_2^2\}^{1/2}} - \Omega(z_1),$$

---

[10] GR 2.264.5.

where we use the symbol Ω here as a shorthand notation that means a second term similar to the first, but with L replaced by $-L$, i.e., the other end of the solenoid. Then we have

$$B_z = \frac{\mu_0 I' a}{2\pi} \int_0^{\pi} \frac{a - \rho \cos \phi'}{(a^2 + \rho^2 - 2a\rho \cos \phi')} \frac{z_2}{\left\{a^2 + \rho^2 + z_2^2 - 2a\rho \cos \phi'\right\}^{1/2}} d\phi' - \Omega(z_1).$$

Change the integration variable using

$$\cos \phi' = -1 + 2x^2$$
$$d\phi' = -\frac{2}{\sqrt{1 - x^2}} dx. \tag{7.38}$$

This gives

$$B_z = \frac{\mu_0 I' a}{\pi} \int_0^1 \frac{a + \rho - 2\rho x^2}{[(a+\rho)^2 - 4a\rho x^2]} \frac{z_2}{\left\{(a+\rho)^2 + z_2^2 - 4a\rho x^2\right\}^{1/2} \sqrt{1 - x^2}} dx - \Omega(z_1).$$

We can put the integral into a standard form by defining

$$k^2 = \frac{4a\rho}{(a+\rho)^2 + z_2^2} \tag{7.39}$$

and

$$n = \frac{4a\rho}{(a+\rho)^2}. \tag{7.40}$$

We find that

$$B_z = \frac{\mu_0 I' a}{\pi (a+\rho)^2} \frac{z_2}{\left\{(a+\rho)^2 + z_2^2\right\}^{1/2}} [(a+\rho) \, \mathbb{I}_2 - 2\rho \, \mathbb{I}_3] - \Omega(z_1),$$

where[11]

$$\mathbb{I}_2 = \int_0^1 \frac{dx}{(1 - n x^2)\{(1 - k^2 x^2)\,(1 - x^2)\}^{1/2}}$$
$$= \Pi(k, -n).$$

The function $\Pi(k, n)$ is the complete elliptic integral of the third kind. The other integral is

---

[11] GR 8.111.4.

$$\mathbb{I}_3 = \int_0^1 \frac{x^2}{(1 - n x^2)\{(1 - k^2 x^2)(1 - x^2)\}^{1/2}} \, dx.$$

This can be evaluated by writing it in the form

$$\mathbb{I}_3 = \frac{1}{n} \int_0^1 \frac{1 - nx^2 - 1}{(1 - n x^2)\{(1 - k^2 x^2)(1 - x^2)\}^{1/2}} \, dx$$

$$= \frac{1}{n} K(k) - \frac{1}{n} \Pi(k, n),$$

where[12]

$$K(k) = \int_0^1 \frac{dx}{\{(1 - k^2 x^2)(1 - x^2)\}^{1/2}}. \tag{7.41}$$

Substituting, we find that the axial field of the solenoid is [22]

$$B_z = \frac{\mu_0 I'}{\pi} \frac{a z_2}{(a + \rho)\{(a + \rho)^2 + z_2^2\}^{1/2}} \left[K(k) + \frac{a - \rho}{2a} (\Pi(k, -n) - K(k)) \right] - \Omega(z_1). \tag{7.42}$$

Selecting instead the x component of the cross-product in the Biot-Savart equation, the transverse component of the solenoid field is

$$B_\rho = \frac{\mu_0 I' a}{2\pi} \int_{-L/2}^{L/2} \int_0^\pi \frac{(z - z') \cos \phi'}{\{a^2 + \rho^2 - 2a\rho \cos \phi' + (z - z')^2\}^{3/2}} \, dz' \, d\phi'$$

$$= \frac{\mu_0 I' a}{2\pi} \int_0^\pi \cos \phi' [z \, \mathbb{I}_1 - \mathbb{I}_4] \, d\phi'.$$

The integral over z' involves the integral $\mathbb{I}_1$ that we have already considered and the integral[13]

$$\mathbb{I}_4 = \int_{-L/2}^{L/2} \frac{z'}{\{e - 2 z z' + z'^2\}^{3/2}} \, dz'$$

$$= \frac{e - \dfrac{z L}{2}}{(e - z^2)\left\{ a^2 + \rho^2 - 2 \, a\rho \cos \phi' + \left(\dfrac{L}{2} - z \right)^2 \right\}^{1/2}} - \Omega(-L),$$

---

[12] GR 8.111.2 and 8.112.1.   [13] GR 2.264.6.

where e was defined in Equation 7.37. Substituting these back into the equation for B_ρ and simplifying, we get

$$B_\rho = \frac{\mu_0 I' a}{2\pi} \int_0^\pi \frac{\cos \phi'}{\{a^2 + \rho^2 + z_2^2 - 2a\rho \cos \phi'\}^{1/2}} d\phi' - \Omega(z_1).$$

Making the change of variable in Equation 7.38, we find

$$B_\rho = \frac{\mu_0 I' a}{\pi} \frac{1}{\{(a+\rho)^2 + z_2^2\}^{1/2}} \int_0^1 \frac{2x^2 - 1}{\{(1 - k^2 x^2)(1 - x^2)\}^{1/2}} dx - \Omega(z_1)$$

$$= \frac{\mu_0 I' a}{\pi} \frac{1}{\{(a+\rho)^2 + z_2^2\}^{1/2}} [2\,\mathbb{I}_5 - K(k)] - \Omega(z_1),$$

where k^2 was defined in Equation 7.39. The remaining integral is[14]

$$\mathbb{I}_5 = \int_0^1 \frac{x^2}{\{(1 - k^2 x^2)(1 - x^2)\}^{1/2}} dx$$

$$= \frac{K(k) - E(k)}{k^2}.$$

Substituting and simplifying, we find the transverse component of the solenoid field is [22]

$$B_\rho = \frac{\mu_0 I'}{4\pi} \frac{\{(a+\rho)^2 + z_2^2\}^{1/2}}{\rho} [2(K(k) - E(k)) - k^2 K(k)] - \Omega(z_1). \quad (7.43)$$

Equations 7.42 and 7.43 are exact solutions for the sheet solenoid that are valid for all points in space, except for observation points at the same radius as the current sheet. For an arbitrary field point with the cylindrical coordinates (ρ, ϕ, z), we can write the Cartesian values of the transverse field as

$$B_x = B_\rho \cos \phi = B_\rho \frac{x}{\rho}$$

$$B_y = B_\rho \sin \phi = B_\rho \frac{y}{\rho}. \quad (7.44)$$

An alternate solution for the field components has also been given in terms of related functions known as generalized complete elliptic integrals.[23]

The vector potential for the sheet solenoid may also be expressed exactly in terms of elliptic integrals [22, 24] as

---

[14] GR 3.153.5.

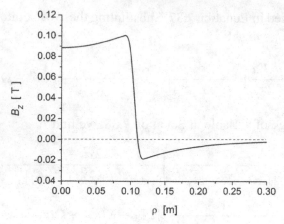

Figure 7.10 The dependence of B_z on ρ at the center of a solenoid with $L = 20$ cm, $a = 10$ cm, $I' = 10^5$ A/m.

$$A_\phi = \frac{\mu_0 I'}{4\pi} \frac{z_2}{\rho} \left[\{(a+\rho)^2 + z_2^2\}^{1/2}(K(k) - E(k)) \right.$$

$$\left. - \frac{(a-\rho)^2}{\{(a+\rho)^2 + z_2^2\}^{1/2}} \ (\Pi(k,n) - K(k)) \right] - \Omega(z_1), \qquad (7.45)$$

where k and n are given by Equations 7.39 and 7.40, respectively. The vector potential and field for the sheet solenoid may also be expressed as sums of Bessel-Laplace integrals [25, 26] or in terms of modified Bessel functions.[27]

Example 7.2:   radial dependence of the axial solenoid field
Let us examine the dependence of B_z on ρ at the center of a sample solenoid. The results from using Equation 7.42 are shown in Figure 7.10. Note the reversal of the field direction at the radius of the sheet.

Example 7.3:   mutual inductance and axial force between a solenoid and a loop
Once the vector potential and the field components for the circular current loop and the sheet solenoid are known, the mutual inductance between a solenoid and a current loop can be computed as

$$M(S, L) = \frac{\Phi_L}{I_S} = \frac{1}{I_S} \int A_\phi(S) \, dl$$

$$= \frac{2\pi \rho_L}{I_S} \ A_\phi(S),$$

where the symbols S and L refer to the solenoid and the loop.

Assume the current is flowing in the same direction in the loop and in the solenoid. The axial force acting on a current loop due to the magnetic field from the solenoid is

$$F_z(S,L) = I_L \int \overrightarrow{dl_L} \times \overrightarrow{B}(S) = -I_L \int_0^{2\pi} \rho_L B_\rho(S) \ d\phi$$
$$= -2\pi I_L \rho_L B_\rho(S),$$

where the minus sign indicates that the force tries to pull the loop and the solenoid together.

7.6 Block model for the solenoid

In cases where the accuracy of the field calculated from the sheet model for the solenoid is inadequate, it may be necessary to take into account the radial thickness of the coils. Consider the cross-section of a block solenoid shown in Figure 7.11, where the coil extends from an inner radius a to an outer radius b. We can find the on-axis axial field by integrating Equation 7.31 for the field due to a current sheet

$$B_z(0,z) = \frac{\mu_0 J}{2} \int_a^b \frac{z + L/2}{\left\{r^2 + (z + L/2)^2\right\}^{1/2}} \ dr - \Omega(-L),$$

where again Ω is used as a shorthand for the expression in the first term with L replaced by $-L$. Performing the integral,[15] we find for points along the symmetry axis

$$B_z(0,z) = \frac{\mu_0 J}{2} \left\{ (z + L/2) \ \ln\left[\frac{b + \left\{b^2 + (z + L/2)^2\right\}^{1/2}}{a + \left\{a^2 + (z + L/2)^2\right\}^{1/2}} \right] \right\} - \Omega(-L). \quad (7.46)$$

The axial field at the center of the solenoid is

$$B_z(0,0) = \frac{\mu_0 J L}{2} \ \ln\left[\frac{b + \left\{b^2 + L^2/4\right\}^{1/2}}{a + \left\{a^2 + L^2/4\right\}^{1/2}} \right]. \quad (7.47)$$

The off-axis field from the block solenoid is usually treated by summing over the fields from a set of current sheets, using one of the methods we have previously discussed. For example, the block conductor may be simulated using a radial distribution of current sheets expressed in elliptic integrals.[22] It is also possible to express the thick solenoid field in terms of a radial expansion of the on-axis field, [9] as a series of zonal harmonics,[10, 11, 21] or in terms of Bessel-Laplace integrals.

---

[15] GR 2.271.5.

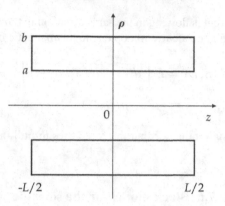

Figure 7.11 Block model of a solenoid.

[25, 26, 28] The good field region calculated for a solenoid from a properly designed block conductor is frequently larger than that for a sheet solenoid.[21]

The flux leaving a solenoid travels outside and returns through the opposite end. As a result, the fringe field on the outside of the solenoid can be quite significant. If the fringe field is unacceptable, it can be reduced by adding supplemental bucking coils or by using iron shielding. Figure 7.12 shows a POISSON model for the magnetic field in a typical solenoid. The figure shows 1/4 of a plane projection through the solenoid. The vertical axis is the centerline of the solenoid. The field is symmetric on both sides of the vertical axis and both sides of the horizontal axis. The program used Dirichlet boundary conditions on the left, right, and top borders, and Neumann boundary conditions on the bottom border. The figure on the left shows the field from just the coil, while the figure on the right illustrates the reduction in the exterior field from adding a cylindrical iron return yoke. The current in the coil was the same for both figures.

7.7 Bent solenoid

So far we have been considering configurations where current loops or solenoids have been symmetrically configured along a straight axis. In the cylindrical coordinate system we have been using, the current has been azimuthally symmetric along ϕ, the system axis has been along z, and the magnetic field only has components along ρ and z. We now generalize this to consider configurations where a solenoid, for example, is bent to follow a circular axis. The magnetic field of a bent solenoid channel is conveniently defined in terms of a rotating coordinate system that follows some reference curve, as shown in Figure 7.13. In the curvilinear description of orthogonal coordinate systems,[29] changes in the values of the coordinates (u_1, u_2, u_3) are related to the distance element by

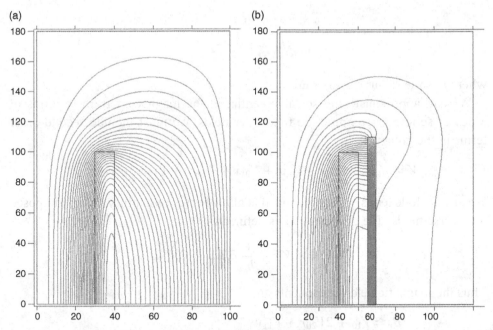

Figure 7.12 Magnetic field of a solenoid coil (left); field for a coil surrounded by an iron return path (right).

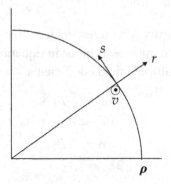

Figure 7.13 Frenet-Serret coordinate system.

$$ds^2 = h_1^2 u_1^2 + h_2^2 u_2^2 + h_3^2 u_3^2,$$

where (h_1, h_2, h_3) are a set of scale factors. In the Frenet-Serret coordinate system considered here,[30, 31] the reference curve is a circle and the origin of the unit vectors moves along the circle. The unit vector s is in the bending plane and tangent to the circle. The unit vector r is in the bending plane and perpendicular to s. The unit vector v is perpendicular to the bending plane. The curvilinear scale factors are

$$h_r = h_v = 1$$

$$h_s = 1 \; + \frac{r}{\rho},$$

where ρ is the radius of curvature.

We can approximate the scalar potential in the magnet aperture in terms of a power series. To compute the first-order fields, we must include second-order terms in the potential,

$$V = \mu_0 V_m \simeq V_{00} + V_{10}r + V_{01}v + V_{20}r^2 + V_{11}rv + V_{02}v^2 \; .$$

We also include terms in the potential that allow for the possibility of superimposed transverse fields. The gradient of V is defined as

$$\nabla V = \partial_r V \, \hat{r} + \frac{1}{h_s} \; \partial_s V \, \hat{s} + \partial_v V \, \hat{v}.$$

Thus the magnetic field components are

$$-B_r \simeq V_{10} + 2V_{20}r + V_{11}v$$

$$-B_v \simeq V_{01} + V_{11}r + 2V_{02}v$$

$$-B_s \simeq \frac{1}{h_s}(V'_{00} + V'_{10}r + V'_{01}v + V'_{20}r^2 + V'_{11}rv + V'_{02}v^2),$$

where primes indicate derivatives with respect to s. Recalling the midplane expansions of the transverse field components given in Equation 4.9, we can associate the potential terms with the multipole field coefficients[16]

$$V_{01} = -B_1$$

$$V_{10} = A_1$$

$$V_{11} = -B_2$$

$$2V_{20} = A_2.$$

Thus to the first-order, the field components are

$$B_r \simeq \; - A_1 - A_2\,r + B_2\,v$$

$$B_v \simeq B_1 + B_2\,r - 2 \; V_{02}\,v \tag{7.48}$$

$$B_s \simeq \frac{1}{h_s} \; (b_s - A'_1\,r + B'_1\,v),$$

---

[16] In the case where a charged particle has to follow the reference path in the horizontal plane, we must have the horizontal dipole field $A_1(s) = 0$.

where we identify the on-axis component of the axial field as

$$b_s = -V_{00}'$$

and assume the transverse dipole fields can vary with s.

At this point, we still have one unidentified potential term V_{02} in Equation 7.48, so we demand that the field components also satisfy the divergence relation $\nabla \cdot \vec{B} = 0$. In the coordinates discussed here, this can be written as

$$\frac{1}{h_s} \partial_r(h_s B_r) + \frac{1}{h_s} \partial_s B_s + \partial_v B_v = 0.$$

Inserting the field components from Equation 7.48 and using

$$\frac{1}{h_s} \simeq 1 - \frac{r}{\rho},$$

we find the constraint

$$-A_2 - \frac{1}{\rho} A_1 - 2 \, V_{02} + b_s' = 0.$$

Thus the first-order vertical field component is

$$B_v \simeq B_1 + B_2 \, r - \left(b_s' - \frac{A_1}{\rho} - A_2 \right) v. \tag{7.49}$$

If no superimposed transverse fields are present, the first-order axial field in the bent channel is

$$B_s \simeq b_s - \frac{r}{\rho} b_s. \tag{7.50}$$

It is also possible to define an expansion for the field that makes use of "curved multipoles" that directly correspond to the solution of Laplace's equation in the curved coordinate system.[32]

7.8 Toroid

When the bent solenoid channel is extended to form a closed circular ring, we have a *toroid*, as shown in Figure 7.14. The current loops from the solenoid are centered on the circular system axis and the plane of the loops lie in the ρ-z plane. The direction of the unit vectors $\hat{\rho}$ and $\hat{\phi}$ depend on the azimuthal location around the toroid. Since the coils are closer together on the side nearer to the center of

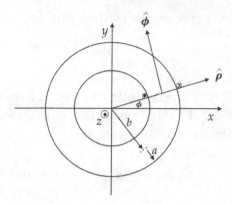

Figure 7.14 Geometry of the toroid from above.

curvature, we expect that the field inside the bent solenoid will have a gradient with respect to the system axis, in agreement with Equation 7.50.

Because of the symmetry of the configuration, all of the field components must be independent of the azimuthal angle ϕ. Let the mean radius of the toroid equal b and the radius of the current loops equal a. Then a simple application of the Ampère law on the midplane ($z = 0$) shows that $B_\phi = 0$ for $\rho < b - a$ and for $\rho > b + a$ since no net current is enclosed in a circular path in those regions. However, applying the Ampère law on a circular path on the midplane, we find the field inside the toroid is

$$B_\phi = \frac{\mu_0 N I}{2\pi\rho},\tag{7.51}$$

where N is the number of conductor turns around the circumference and ρ is the radius of the path. This shows that the field varies like $1/\rho$ inside the toroid.

A cross-section of the toroid at some azimuthal angle ϕ is shown in Figure 7.15a. The angle α gives the location of an element of the current loop. The current element has the Cartesian coordinates

$$x_l = (b + a \cos \alpha) \cos \phi$$
$$y_l = (b + a \cos \alpha) \sin \phi$$
$$z_l = a \sin \alpha$$

and the directions

$$dl_x = a \sin \alpha \cos \phi \, d\alpha$$
$$dl_y = a \sin \alpha \sin \phi \, d\alpha$$
$$dl_z = -a \cos \alpha \, d\alpha.$$

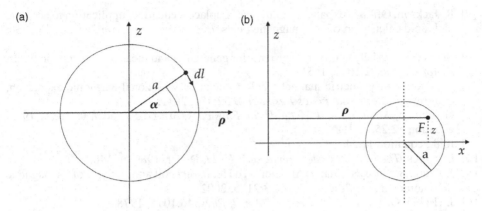

Figure 7.15 (a) Cross-section of the toroid at the azimuthal location ϕ; (b) location of the field observation point F.

The location of the field observation point $(\rho, 0, z)$ is shown in Figure 7.15b. The resulting distance vector is

$$\vec{R} = [\rho - (b + a\cos\alpha)\cos\phi]\,\hat{x} - (b + a\cos\alpha)\sin\phi\,\hat{y} + (z - a\sin\alpha)\,\hat{z}.$$

Applying the Biot-Savart law to any point inside the toroid shows that B_z and B_ρ vanish. It follows that the field inside the toroid has to have the form

$$B = B_\phi(\rho, z).$$

The analytic results for B_ϕ are complicated [33] expressions defined in terms of integrals of elliptic integrals. Alternatively, one could examine the field inside the toroid by evaluating one of the integrals in terms of complete elliptic integrals and performing the other integral numerically.

References

[1] W. Smythe, *Static and Dynamic Electricity*, 2nd ed., McGraw-Hill, 1950, p. 270–275.
[2] J. Stratton, *Electromagnetic Theory*, McGraw-Hill, 1941, p. 263.
[3] R. Good, Elliptic integrals, the forgotten functions, *Eur. J. Phys.* 22:119, 2001.
[4] W. Panofsky & M. Phillips, *Classical Electricity and Magnetism*, 2nd ed., Addison-Wesley, 1962, p. 156.
[5] R. Schill, General relation for the vector magnetic field of a circular current loop: a closer look, *IEEE Trans. Magnetics* 39:961, 2003.
[6] J. D. Jackson, *Classical Electrodynamics*, Wiley, 1962, p. 142.
[7] D. Redzic, The magnetic field of a static current loop: a new derivation, *Eur. J. Phys.* 27:N9, 2006.
[8] G. Harnwell, *Principles of Electricity and Magnetism*, 2nd ed., McGraw-Hill, 1949, p. 329.

[9] R. Jackson, Off-axis expansion solution of Laplace's equation: application to accurate and rapid calculation of coil magnetic fields, *IEEE Trans. Electron Devices* 46:1050, 1999.

[10] M. Garrett, Axially symmetric systems for generating and measuring magnetic fields, *J. Appl. Phys.* 22:1091, 1951.

[11] F. Gluck, Axisymmetric magnetic field calculation with zonal harmonic expansion, *Progress in Electromagnetics Research B* 32:351, 2011.

[12] G. Arfken, *Mathematical Methods for Physicists*, 3rd ed., Academic Press, 1985, equation 12.25.

[13] Ibid., equation 12.24.

[14] L. Eyges, *The Classical Electromagnetic Field*, Dover, 1980, p. 140.

[15] J. Wang, S. She & S. Zhang, An improved Helmholtz coil and analysis of its magnetic field homogeneity, *Rev. Sci. Inst.* 73:2175, 2002.

[16] J. Higbie, Off-axis Helmholtz field, *Am. J. Phys.* 46:1075, 1978.

[17] E. Purcell, Helmholtz coils revisited, *Am. J. Phys.* 57:18, 1989.

[18] P. Murgatroyd, Optimum designs of circular coil systems for generating non-uniform axial magnetic fields, *Rev. Sci. Inst.* 50:668, 1979.

[19] P. Murgatroyd & B. Bernard, Inverse Helmholtz pairs, *Rev. Sci. Inst.* 54:1736, 1983.

[20] R. Merritt, C. Purcell & G. Stroink, Uniform magnetic field produced by three, four and five square coils, *Rev. Sci. Inst.* 54:879, 1983.

[21] M. Garrett, Thick cylindrical coil systems for strong magnetic fields with field or gradient homogeneities of the 6th to 20th order, *J. Appl. Phys.* 38:2563, 1967.

[22] M. Garrett, Calculation of fields, forces and mutual inductances of current systems by elliptic integrals, *J. Appl. Phys.* 34:2567, 1963.

[23] N. Derby & S. Olbert, Cylindrical magnets and ideal solenoids, *Am. J. Phys.* 78:229, 2010.

[24] L. Cohen, An exact formula for the mutual inductance of coaxial solenoids, *Bulletin Bureau Standards* 3:295, 1907.

[25] J. Conway, Exact solutions for the magnetic fields of axisymmetric solenoids and current distributions, *IEEE Trans. Mag.* 37:2977, 2001.

[26] J. Conway, Trigonometric integrals for the magnetic field of a coil of rectangular cross section, *IEEE Trans. Mag.* 42:1538, 2006.

[27] T. Tominaka, Magnetic field calculation of an infinitely long solenoid, *Eur. J. Phys.* 27:1399, 2006.

[28] V. Labinac, N. Erceg & D. Kotnik-Karuza, Magnetic field of a cylindrical coil, *Am. J. Phys.* 74:621, 2006.

[29] W. Panofsky & M. Phillips, *op. cit.*, p. 473–476.

[30] S. Y. Lee, *Accelerator Physics*, World Scientific, 1999, p. 33–34.

[31] C. Wang & L. Teng, Magnetic field expansion for particle tracking in a bent solenoid channel, Proc. 2001 *Particle Accelerator Conference*, p. 456. Available from www .jacow.org.

[32] S. Mane, Solutions of Laplace's equation in two dimensions with a curved longitudinal axis, *Nuc. Instr. Meth. A* 321:365, 1992.

[33] N. Carron, On the fields of a torus and the role of the vector potential, *Am. J. Phys.* 63:717, 1995.

8

Periodic magnetic channels

In the previous chapters, we considered magnets that for the most part had continuous transverse or longitudinal fields along some system axis. In this chapter, we look instead at magnetic channels where the on-axis field is periodic. Periodic field configurations are used for focusing charged particle beams and for production of radiation at light sources. We begin by considering the field produced by helical conductor windings. Then we examine several examples where we demand some desired field configuration along the axis and then find off-axis field components that satisfy Maxwell's equations.

8.1 Field from a helical conductor

A helically wound conductor can produce a periodic field. The parametric equations of a helix are

$$x = a \cos \phi$$
$$y = a \sin \phi$$
$$z = \frac{\lambda}{2\pi} \phi,$$

where a is the radius and λ is the period of the helix. We define the axial wavenumber as $k = 2\pi/\lambda$. We parameterize the nature of the helix by the angle α in Figure 8.1.[1] We have

$$\tan \alpha = \frac{\Delta z}{a \, \Delta \phi} = \frac{\lambda}{2\pi a} = \frac{1}{ka}$$

since z progresses by a distance λ as the azimuthal angle goes once around the circumference. With this definition, $\alpha = 0$ corresponds to the limiting case when the helix reduces to a circular loop. It is convenient to write λ as a function of a.

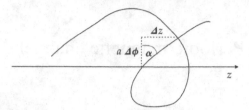

Figure 8.1 Definition of the helical angle α.

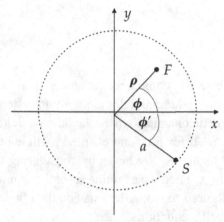

Figure 8.2 Helical conductor geometry at a fixed value of z.

$$\lambda = 2\pi a \tan \alpha$$

The azimuthal angle and axial distance are connected through the helix constraint

$$z = a \phi \tan \alpha. \tag{8.1}$$

Consider the cross-section through the helix shown in Figure 8.2. A single conductor follows a helical path in a current sheet of radius a. Let the observation point F have cylindrical coordinates (ρ, ϕ, z) and the current element at the location S on the conductor have coordinates (a, ϕ', z'). The current element is given by

$$\vec{dl} = -a \sin \phi' \, d\phi' \, \hat{x} + a \cos \phi' \, d\phi' \, \hat{y} + a \tan \phi' \, d\phi' \, \hat{z}$$

and the distance vector is

$$\vec{R} = (\rho \cos \phi - a \cos \phi') \, \hat{x} + (\rho \sin \phi - a \sin \phi') \, \hat{y} + (z - a \phi' \tan \alpha) \, \hat{z}.$$

Taking into account the constraint between z' and ϕ', the direct evaluation of $\boldsymbol{B}$ using the Biot-Savart equation only requires an integration over ϕ'. The integration limits for a winding of finite length can be found using Equation 8.1. Although the resulting integral for the general case is complicated, the solution for observation points along the axis of the helix is fairly straightforward.[1]

The vector potential for an infinitely long helix is given by

$$\overrightarrow{A}(\rho,\phi,z) = \frac{\mu_0}{4\pi} \int \frac{\overrightarrow{K}(a,\phi',z')}{R} \, dS'. \tag{8.2}$$

The sheet current density only has components in the ϕ' and z' directions. The pitch angle α for the helical winding can be written as

$$\tan \alpha = \frac{K_{z'}}{K_{\phi'}},$$

so the components of the current density are related by

$$K_{\phi'} = ka \, K_{z'}.$$

Thus the current density is [2]

$$\overrightarrow{K}(a,\phi',z') = \frac{I}{a} \, (\hat{z}' + ka\hat{\phi}') \, \delta(\phi' - kz' - \varepsilon), \tag{8.3}$$

where ε is the azimuthal angle of the winding at $z' = 0$. The Dirac delta function enforces the constraint between changes in z' and changes in ϕ'.

We can write the periodic delta function in Equation 8.3 as a Fourier series. Let $\tau = \varepsilon + kz'$. Then

$$f(\phi') - \delta(\phi' - \tau) = a_0 + \sum_{n=1}^{\infty} [a_n \cos n\phi' + b_n \sin n\phi'].$$

The coefficients are

$$a_0 = \frac{1}{2\pi} \int_{-\pi}^{\pi} \delta(\phi' - \tau) \, d\phi' = \frac{1}{2\pi}$$

$$a_n = \frac{1}{\pi} \int_{-\pi}^{\pi} \delta(\phi' - \tau) \cos n\phi' \, d\phi' = \frac{1}{\pi} \cos n\tau, \qquad n > 0$$

$$b_n - \frac{1}{\pi} \int_{-\pi}^{\pi} \delta(\phi' - \iota) \, \sin n\phi' \, d\phi' = \frac{1}{\pi} \sin n\iota.$$

Thus the delta function can be expressed as

$$\delta(\phi' - \tau) = \frac{1}{2\pi} + \frac{1}{\pi} \sum_{n=1}^{\infty} [\cos n\tau \cos n\phi' + \sin n\tau \sin n\phi']$$

$$= \frac{1}{2\pi} \left[1 + 2 \sum_{n=1}^{\infty} \cos \left(n(\phi' - \tau) \right) \right]. \tag{8.4}$$

The distance from the current element to the observation point can be written as

$$R = \{a^2 + \rho^2 - 2\, a\rho \cos(\phi - \phi') + (z - z')^2\}^{1/2}. \tag{8.5}$$

We need to express the unit vectors in Equation 8.3 in terms of unit vectors in the coordinate system for the observation point. The axial unit vectors are identical, $\hat{z}' = \hat{z}$. The azimuthal unit vector is given by

$$\hat{\phi}' = -\hat{\rho}\, \sin(\phi - \phi') + \hat{\phi}\, \cos(\phi - \phi'). \tag{8.6}$$

Thus there are in general nonvanishing components of the vector potential in all three dimensions.

Substituting Equations 8.3–8.6 into Equation 8.2, the *axial* component of the vector potential is given by

$$A_z(\rho, \phi, z) = \frac{\mu_0}{4\pi} \frac{I}{2\pi a} \iint \frac{\left[1 + 2 \sum_{n=1}^{\infty} \cos \left(n(\phi' - kz' - \varepsilon) \right) \right]}{\{a^2 + \rho^2 - 2\, a\rho \cos(\phi - \phi') + (z - z')^2\}^{1/2}}\, a\, d\phi'\, dz'. \tag{8.7}$$

For $\rho < a$, evaluation of the integrals give [2]

$$A_z(\rho, \phi, z) = -\frac{\mu_0 I}{2\pi} \ln a + \frac{\mu_0 I}{\pi} \sum_{n=1}^{\infty} K_n(nka) I_n(nk\rho) \cos\left(n(\phi - kz - \varepsilon) \right) \tag{8.8}$$

and for $\rho > a$

$$A_z(\rho, \phi, z) = -\frac{\mu_0 I}{2\pi} \ln \rho + \frac{\mu_0 I}{\pi} \sum_{n=1}^{\infty} K_n(nk\rho) I_n(nka) \cos\left(n(\phi - kz - \varepsilon) \right). \tag{8.9}$$

The functions K_n and I_n are modified Bessel functions of order n.

The *radial* component of the vector potential is given by

$$A_\rho(\rho,\phi,z) = \frac{\mu_0}{4\pi} \frac{I\,k}{2\pi} \iint \frac{\sin(\phi-\phi')\left[1+2\sum\limits_{n=1}^{\infty}\cos\left(n(\phi'-kz'-\varepsilon)\right)\right]}{\{a^2+\rho^2-2\,a\rho\,\cos(\phi-\phi')+(z-z')^2\}^{1/2}}\,a\,d\phi'\,dz'.$$

$$(8.10)$$

For $\rho < a$, evaluation of the integrals give [2]

$$A_\rho(\rho,\phi,z) = -\frac{\mu_0 Ika}{2\pi}\sum_{n=1}^{\infty}[K_{n+1}(nka)\,I_{n+1}(nk\rho) - K_{n-1}(nka)\,I_{n-1}(nk\rho)]$$

$$\sin\left(n(\phi-kz-\varepsilon)\right)$$

$$(8.11)$$

and for $\rho > a$

$$A_\rho(\rho,\phi,z) = -\frac{\mu_0 Ika}{2\pi}\sum_{n=1}^{\infty}[K_{n+1}(nk\rho)\,I_{n+1}(nka) - K_{n-1}(nk\rho)\,I_{n-1}(nka)]$$

$$\sin\left(n(\phi-kz-\varepsilon)\right).$$

$$(8.12)$$

The *azimuthal* component of the vector potential is given by

$$A_\phi(\rho,\phi,z) = \frac{\mu_0}{4\pi}\frac{I\,k}{2\pi}\iint \frac{\cos(\phi-\phi')\left[1+2\sum\limits_{n=1}^{\infty}\cos\left(n(\phi'-kz'-\varepsilon)\right)\right]}{\{a^2+\rho^2-2\,a\rho\,\cos(\phi-\phi')+(z-z')^2\}^{1/2}}\,a\,d\phi'\,dz'.$$

$$(8.13)$$

For $\rho < a$ this has the solution [2]

$$A_\phi(\rho,\phi,z) = \frac{\mu_0 I\,k\rho}{4\pi}$$

$$+\frac{\mu_0 I\,ka}{2\pi}\sum_{n=1}^{\infty}[K_{n+1}(nka)\,I_{n+1}(nk\rho) + K_{n-1}(nka)\,I_{n-1}(nk\rho)]\cos\left(n(\phi-kz-\varepsilon)\right)$$

$$(8.14)$$

and for $\rho > a$

$$A_\phi(\rho,\phi,z) = \frac{\mu_0 I \, ka^2}{4\pi\rho}$$

$$+\frac{\mu_0 I \, ka}{2\pi} \sum_{n=1}^{\infty} [K_{n+1}(nk\rho) \, I_{n+1}(nka) + K_{n-1}(nk\rho) \, I_{n-1}(nka)] \cos\Big(n(\phi - kz - \varepsilon)\Big).$$

$$(8.15)$$

The solution for the magnetic field components can be found by taking the curl of A. For the case $\rho < a$ the field components are [2, 3, 4]

$$B_\rho(\rho,\phi,z) = -\frac{\mu_0 I k^2 a}{\pi} \sum_{n=1}^{\infty} n K_n'(nka) \, I_n'(nk\rho) \sin(n(\phi - kz - \varepsilon))$$

$$B_\phi(\rho,\phi,z) = \frac{\mu_0 I k a}{\pi} \sum_{n=1}^{\infty} n K_n'(nka) \, \frac{I_n(nk\rho)}{\rho} \cos(n(\phi - kz - \varepsilon)) \qquad (8.16)$$

$$B_z(\rho,\phi,z) = \frac{\mu_0 I k}{2\pi} - \frac{\mu_0 I k^2 a}{\pi} \sum_{n=1}^{\infty} n K_n'(nka) \, I_n(nk\rho) \cos(n(\phi - kz - \varepsilon)),$$

while the solution for the case $\rho > a$ is

$$B_\rho(\rho,\phi,z) = -\frac{\mu_0 I k^2 a}{\pi} \sum_{n=1}^{\infty} n K_n'(nk\rho) \, I_n'(nka) \sin(n(\phi - kz - \varepsilon))$$

$$B_\phi(\rho,\phi,z) = \frac{\mu_0 I}{2\pi\rho} + \frac{\mu_0 I k a}{\pi} \sum_{n=1}^{\infty} n I_n'(nka) \, \frac{K_n(nk\rho)}{\rho} \cos(n(\phi - kz - \varepsilon)) \qquad (8.17)$$

$$B_z(\rho,\phi,z) = -\frac{\mu_0 I k^2 a}{\pi} \sum_{n=1}^{\infty} n I_n'(nka) \, K_n(nk\rho) \cos(n(\phi - kz - \varepsilon)).$$

Primes on the Bessel functions indicate derivatives with respect to the arguments.

An example of the variation of the field components for a helical conductor is shown for one period in Figure 8.3. The calculations were done using Equations 8.16. In this case, the magnitude of the transverse components are small compared to the axial component. At radii large compared to a, the azimuthal field component becomes dominant and the field approaches that of a straight wire.

The on-axis field of the helical conductor can be found by evaluating Equation 8.16 at $\rho = 0$. Using the relations

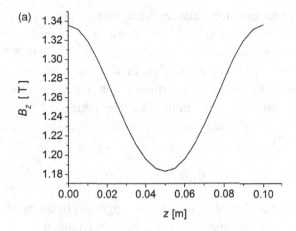

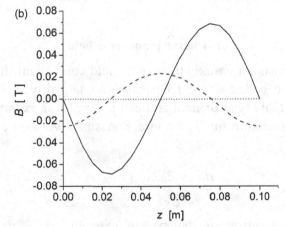

Figure 8.3 (a) The dependence of B_z along one period of the helix; (b) the dependence of B_ρ (solid) and B_ϕ (dashed) versus z. The parameters used here were $\lambda = 10$ cm, $a = 10$ cm, $\rho = 5$ cm, $\phi = 0$, $\varepsilon = 0$, $I = 10^5$ A, and $N = 40$ terms in the sums.

$$I_n(0) = 0 \qquad \text{for } n > 0$$
$$I_n'(0) = 0 \qquad \text{for } n > 1$$
$$I_1'(0) = \tfrac{1}{2},$$

the on-axis field is [1, 3, 5]

$$B_\rho(0,0,z) = \frac{\mu_0 I k^2 a}{2\pi} \, K_1'(ka)$$
$$B_\phi(0,0,z) = 0$$
$$B_z(0,0,z) = \frac{\mu_0 I \, k}{2\pi}.$$

There is a close relationship between these results and our previous results for the field of a solenoid. In practice, the conductor in a solenoid is wound in many helical layers over a cylindrical form. The helical pitch length λ for a solenoid is very small. In the previous chapter, the field for a solenoid was derived by assuming that the field came from a longitudinal distribution of parallel infinitesimal current loops. In the limit that $k \to \infty$, it can be shown by taking the asymptotic limits for the Bessel functions that Equation 8.16 approaches the on-axis field of an infinitely long solenoid

$$B_z(0,0,z) = \mu_0 nI,$$

where $n = 1/\lambda$ is the number of turns per unit length.[6] In the opposite limit where $k \to 0$, the helical fields approach that of a straight conductor.

8.2 Planar transverse field

A planar *wiggler* has an on-axis transverse field component that oscillates in a fixed plane. It is called a wiggler because a charged particle beam moves back and forth in this type of field and can be used, for example, to produce electromagnetic radiation for light sources. Assume we want an on-axis field given by

$$\begin{aligned} B_{y0} &= B_0 \cos\left(\gamma z - \varphi\right) \\ B_{x0} &= B_{z0} = 0, \end{aligned} \tag{8.18}$$

where z is the direction of the system axis and φ is an initial phase. The coefficient γ is related to the wavelength of the field oscillation λ by $\gamma = 2\pi/\lambda$. We saw in Chapter 3 that solutions of Laplace's equation in rectangular coordinates (1) can be written as products of trigonometric and hyperbolic sines and cosines and (2) that these solutions must have at least one trigonometric and one hyperbolic factor. Since B_y is assumed to be non-zero on the axis, we must choose the cosine and hyperbolic cosine functions for the general solution. Once we have specified that the z dependence is a cosine function, there are three possible combinations for the product of the x and y dependences. Let us consider the solution where

$$B_y = B_0 \cos(\alpha x)\,\cosh(\beta y)\,\cos(\gamma z - \varphi).$$

In free space, the div $B = 0$ equation gives

$$\partial_x B_x + B_0\,\beta \cos(\alpha x)\,\sinh(\beta y)\,\cos(\gamma z - \varphi) + \partial_z B_z = 0.$$

In order to satisfy this equation for all x, y, and z, B_x and B_z must have the form

$$B_x = f \, \sin(\alpha x) \, \sinh(\beta y) \, \cos(\gamma z - \varphi)$$
$$B_z = g \, \cos(\alpha x) \, \sinh(\beta y) \, \sin(\gamma z - \varphi),$$

where f and g are unknown factors. Substituting these field expressions back into the divergence equation gives the constraint

$$f\alpha + \beta B_0 + g\gamma = 0. \tag{8.19}$$

The x component of the curl $B = 0$ equation gives the relation

$$g = -\frac{\gamma B_0}{\beta},$$

while either the y or z component of the curl equation gives

$$f = -\frac{\alpha B_0}{\beta}.$$

Substituting these values for f and g into Equation 8.19, we find the wave number constraint

$$\gamma^2 = -\alpha^2 + \beta^2. \tag{8.20}$$

This can be written in terms of the period of the field variation as

$$\lambda^2 = \frac{4\pi^2}{\beta^2 - \alpha^2}.$$

For a periodic solution, we have the additional constraint that $\beta > \alpha$. The solution for the planar transverse field is then [7]

$$B_x = -\frac{\alpha B_0}{\beta} \, \sin(\alpha x) \, \sinh(\beta y) \, \cos(\gamma z - \varphi)$$
$$B_y = B_0 \cos(\alpha x) \, \cosh(\beta y) \, \cos(\gamma z - \varphi) \tag{8.21}$$
$$B_z = -\frac{\gamma B_0}{\beta} \, \cos(\alpha x) \, \sinh(\beta y) \, \sin(\gamma z - \varphi).$$

The other two solutions consistent with Equation 8.18 have the transverse dependences

$$\cosh(\alpha x) \, \cos(\beta y)$$

and

$$\cosh(\alpha x) \cosh(\beta y).$$

These solutions can be derived following the same procedure used above. Each solution has a unique relation among the wavenumbers.[7]

It's important to keep in mind that this type of derivation only represents part of the problem. What we have shown is that our desired on-axis field profile is a valid solution of Maxwell's equations. However, what we have not considered is a configuration of conductors that actually produces that desired field. The obvious trial solution here would be a series of transverse permanent magnet or electrically excited magnetic poles that alternate in direction along the system axis. Oftentimes the field from a realistic coil distribution can only approximate the desired field. We define the problem of finding a current distribution that produces a specified magnetic field as an *inverse problem* to distinguish it from the situation encountered using the Biot-Savart formula, where we find the magnetic field produced by a given current distribution. Finding a suitable current distribution frequently involves using numerical optimization methods.

8.3 Helical transverse field

Consider an on-axis transverse field that rotates around the system axis, analogously to the magnetic field vector in circularly polarized light. It is convenient to look for a solution in a cylindrical coordinate system that rotates around the system axis, as shown in Figure 8.4. In this system, z and ϕ are coupled and the on-axis field is

$$\begin{aligned} B_{\rho 0} &= B_0 \cos(\gamma z - \phi - \varepsilon) \\ B_{\phi 0} &= B_{z0} = 0, \end{aligned} \tag{8.22}$$

where ε is an initial phase shift. For points off the axis, we look for a solution for B_ρ with the form

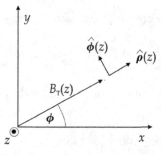

Figure 8.4 Rotating cylindrical coordinate system.

$$B_\rho = C\,F(\rho)\,\cos(\gamma z - \phi - \varepsilon), \tag{8.23}$$

where $CF(0) = B_0$ and F is an undetermined function of ρ. Since the ρ and ϕ coordinates are separated by $90°$, we expect the solution for B_ϕ to have the form

$$B_\phi = C\,G(\rho)\,\sin(\gamma z - \phi - \varepsilon), \tag{8.24}$$

where G is another undetermined function. We know from Chapter 3 that the solution of Laplace's equation in cylindrical coordinates must involve Bessel functions. Calculating the ρ component of the curl $B = 0$ equation allows us to obtain an expression for B_z.

$$B_z = -C\,\gamma\,\rho\,G(\rho)\,\sin(\gamma z - \phi - \varepsilon). \tag{8.25}$$

Calculating the ϕ component of the curl equation lets us determine a relation between the unknown functions F and G.

$$F(\rho) = \partial_\rho[\rho\,G(\rho)]. \tag{8.26}$$

Using Equations 8.23–8.26, we can write the div $B = 0$ equation directly in terms of G.

$$\frac{1}{\rho}[\rho\,\partial_\rho^2(\rho G) + \partial_\rho(\rho G)] - \frac{1}{\rho}\,G - \gamma^2\rho\,G = 0. \tag{8.27}$$

Rearranging terms, this can be written as

$$\gamma^2\rho^2\,\frac{\partial^2(\gamma\,\rho\,G)}{\partial(\gamma\,\rho)^2} + \gamma\,\rho\,\frac{\partial(\gamma\,\rho\,G)}{\partial(\gamma\,\rho)} - [1 + (\gamma\,\rho)^2](\gamma\,\rho\,G) = 0. \tag{8.28}$$

This is the differential equation for the modified Bessel[1] function I_1, where the unknown variable is $\gamma\,\rho G$ and the argument of the Bessel function is $\gamma\,\rho$. Thus we have

$$\gamma\,\rho\,G(\rho) = I_1(\gamma\,\rho)$$

and the unknown function G is

$$G(\rho) = \frac{I_1(\gamma\,\rho)}{\gamma\,\rho}. \tag{8.29}$$

---

[1] Some properties of the modified Bessel functions are described in Appendix C.

We can now find F from Equation 8.26.

$$F(\rho) = \frac{\partial I_1(\gamma\rho)}{\partial(\gamma\rho)}$$

Using a recursion relation,[2] we can write this as

$$F(\rho) = I_0(\gamma\rho) - \frac{1}{\gamma\rho} I_1(\gamma\rho), \tag{8.30}$$

where I_0 is the modified Bessel function of order 0. Substituting this back into Equation 8.23, we find

$$B_\rho = C\left[I_0(\gamma\rho) - \frac{1}{\gamma\rho} I_1(\gamma\rho)\right] \cos(\gamma z - \phi - \varepsilon).$$

Near the axis, I_0 and I_1 have the series expansions[3]

$$I_0(u) \simeq 1 + \frac{1}{4} u^2 + \cdots \tag{8.31}$$

and[4]

$$I_1(u) \simeq \frac{1}{2} u + \frac{1}{16} u^3 + \cdots \tag{8.32}$$

Thus near the axis, we take the leading terms for I_0 and I_1 and find that

$$B_\rho \simeq \frac{C}{2} \cos(\gamma z - \phi - \varepsilon).$$

Comparing this with Equation 8.23 gives $C = 2B_0$. The solution for the helical transverse field is then [8]

$$\begin{aligned}
B_\rho &= 2B_0\left[I_0(\gamma\rho) - \frac{1}{\gamma\rho} I_1(\gamma\rho)\right] \cos(\gamma z - \phi - \varepsilon) \\
B_\phi &= 2B_0 \frac{I_1(\gamma\rho)}{\gamma\rho} \sin(\gamma z - \phi - \varepsilon) \\
B_z &= -2B_0 I_1(\gamma\rho) \sin(\gamma z - \phi - \varepsilon).
\end{aligned} \tag{8.33}$$

Note that the argument for the Bessel functions involves the longitudinal wave-number γ.

---

[2] AS 9.6.26.   [3] AS 9.6.12.   [4] AS 9.6.10.

It was suggested that this solution could be produced by winding conductors in a helical manner over a cylindrical bore tube.[8] Between adjacent helical turns, a second helix could be laid down with the current running in the opposite direction. However, exact calculations showed that Equation 8.33 is only a reasonable approximation for the field of this interleaved helical winding configuration when a/λ is larger than ~0.3, where a is the winding radius.[3, 5]

8.4 Axial fields

As a final example of constructing a desired on-axis field, we consider the periodic axial field

$$B_{z0} = B_0 \cos(\gamma z - \varepsilon)$$
$$B_{\phi 0} = B_{\rho 0} = 0$$

in a cylindrical coordinate system. By symmetry, B_ϕ vanishes everywhere. Because of the periodic behavior in z, we suspect that the off-axis solution must contain terms proportional to the modified Bessel functions. Therefore we look for the simplest possible solution

$$B_z = B_0 \, I_0(\alpha\rho) \cos(\gamma z - \varepsilon).$$

Since the div $B = 0$ equation involves $\partial_z B_z$, we know that B_ρ must be proportional to $\sin(\gamma z - \varepsilon)$. Thus we have

$$\frac{1}{\rho} \, \partial_\rho(\rho B_\rho) = B_0 \, \gamma I_0(\alpha \rho) \sin(\gamma z - \varepsilon).$$

Multiplying by ρ and integrating both sides, we get

$$\rho B_\rho = B_0 \gamma \sin(\gamma z - \varepsilon) \int_0^\rho \rho \, I_0(\alpha\rho) \, d\rho + C,$$

where C is an integration constant. The remaining integral can be evaluated as[5]

$$\int_0^\rho \rho \, I_0(\alpha\rho) \, d\rho = \frac{\rho}{\alpha} \, I_1(\alpha\rho).$$

Thus

$$B_\rho = B_0 \, \frac{\gamma}{\alpha} \, I_1(\alpha\rho) \sin(\gamma z - \varepsilon) + \frac{C}{\rho}.$$

---

[5] AS 11.3.25.

Since B_ρ must vanish at $\rho = 0$, we must have $C = 0$. To find the relationship between the wavenumbers α and γ, we look at the ϕ component of the curl $B = 0$ equation, which gives

$$\frac{\gamma^2}{\alpha} I_1(\alpha\rho) - \frac{\partial I_0(\alpha\rho)}{\partial\rho} = 0.$$

Using[6]

$$\frac{\partial I_0(\alpha\rho)}{\partial\rho} = \alpha\, I_1(\alpha\rho),$$

we find that $\alpha = \gamma$. Thus the solution for the periodic axial field is

$$
\begin{aligned}
B_\rho &= B_0\, I_1(\gamma\rho)\, \sin(\gamma z - \varepsilon) \\
B_\phi &= 0 \\
B_z &= B_0\, I_0(\gamma\rho)\, \cos(\gamma z - \varepsilon).
\end{aligned}
\tag{8.34}
$$

A likely conductor configuration for producing this field is a series of solenoids along the axis that alternate in direction. In fact, it is possible to design block solenoids that give an excellent approximation to a sinusoidal axial field along the axis.[9]

References

[1] W. Smythe, *Static and Dynamic Electricity*, 2nd ed., McGraw-Hill, 1950, p. 276–278.
[2] T. Tominaka, Vector potential for a single helical current conductor, *Nuc. Instr. Meth. A.* 523:1, 2004.
[3] T. Tominaka, M. Okamura & T. Katayama, Analytic field calculation of helical coils, *Nuc. Instr. Meth. A.* 459:398, 2001.
[4] R. Hagel, L. Gong & R. Unbehauen, On the magnetic field of an infinitely long helical line current, *IEEE Trans. Mag.* 30:80, 1994.
[5] S. Park, J. Baird, R. Smith & J. Hirshfield, Exact magnetic field of a helical wiggler, *J. Appl. Phys.* 53:1320, 1982.
[6] T. Tominaka, Magnetic field calculation of an infinitely long solenoid, *Eur. J. Phys.* 27:1399, 2006.
[7] D. Sagan, J. Crittenden, D. Rubin & E. Forest, A magnetic field model for wigglers and undulators, Proc. 2003 Part. Accel. Conf., Portland, OR, p. 1023.
[8] J. Blewett & R. Chasman, Orbits and fields in the helical wiggler, *J. Appl. Phys.* 48:2692, 1977.
[9] R. Fernow & R. Palmer, Solenoidal ionization cooling lattices, *Phys. Rev. Special Topics – Accel. and Beams* 10:064001, 2007.

[6] AS 9.6.27.

9

Permanent magnets

Permanent magnets are ferromagnetic materials that retain significant magnetization after the magnetizing current is removed.[1] In applications where space is constrained, permanent magnets can provide stronger fields than electromagnets, require no power source, and do not need cooling. After discussing the properties of bar magnets and magnetic circuits, we consider some models of the properties of magnets made from rare earth compounds. We conclude with a discussion of assemblies of permanent magnets, which can be used to produce desired multipole fields.

9.1 Bar magnets

Let us consider a cylindrical sample of permanent magnet material with uniform magnetization M along the axis of the cylinder, as shown in Figure 9.1. Recall from Equation 3.23 that the vector potential can be written as

$$\vec{A} = \frac{\mu_0}{4\pi} \int \frac{\vec{M} \times \hat{n}}{R} \, dS + \frac{\mu_0}{4\pi} \int \frac{\nabla \times \vec{M}}{R} \, dV.$$

Since M is uniform here, the second term vanishes. In addition, the first term vanishes on the flat end faces. Thus A only depends on the surface contributions around the sides of the cylinder. We assume the sources for the potential are azimuthally directed Amperian currents with current density

$$
\begin{aligned}
\vec{K}_a &= \vec{M} \times \hat{n} \\
&= M \, \hat{\phi}.
\end{aligned}
\tag{9.1}
$$

Thus the field in a bar magnet is analogous to the field in a solenoid, where the Amperian currents here take the place of the conduction currents in a solenoid. The magnetic flux density in the axial direction is then given by Equation 7.32

211

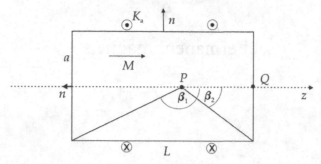

Figure 9.1 Bar magnet.

$$B_z(0, z) = \frac{\mu_0 n I}{2} \left(\cos \beta_2 - \cos \beta_1 \right),$$

where β_i are the angles from the observation point along z to the two outer edges at the ends of the cylinder. Making the substitution $K_a = n I$, we get

$$B = \frac{\mu_0 K_a}{2} \left(\cos \beta_2 - \cos \beta_1 \right)$$

$$= \frac{\mu_0 M}{2} \left(\cos \beta_2 - \cos \beta_1 \right) \tag{9.2}$$

$$= \frac{B_R}{2} \left(\cos \beta_2 - \cos \beta_1 \right),$$

where B_R is the remanent field for the permanent magnet. For a point P inside the magnet, $\cos \beta_1 < 0$ and $\cos \beta_2 > 0$. Thus B points along the positive z direction. From Gauss's law for a pillbox on one of the end faces, B must also be directed along $+z$ outside the magnet. We can find the magnetic intensity H from

$$\mu_0 H = B - \mu_0 M.$$

Thus [2]

$$H_P = \frac{K_a}{2} \left(\cos \beta_2 - \cos \beta_1 \right) - K_a$$

$$= \frac{K_a}{2} \left(\cos \beta_2 - \cos \beta_1 - 2 \right). \tag{9.3}$$

Since the two cosine terms are both smaller than one, this expression is negative. Thus H inside the magnet points in the opposite direction from M and B.

Now let us consider the situation at the point Q on the end face of the magnet. In this case $\cos \beta_2 = 0$ and

$$B_Q = -\frac{\mu_0 K_a}{2} \cos \beta_1. \tag{9.4}$$

On the inside surface of the end face, we find from Equation 9.3 that

$$H_Q^{in} = -\frac{K_a}{2}(\cos \beta_1 + 2),$$

which points along $-z$. On the outer surface of the end face, $M = 0$ and we find from Equation 9.4 that

$$H_Q^{out} = -\frac{K_a}{2} \cos \beta_1,$$

which points along $+z$. The behavior for H is similar to the case of the electric field from a surface distribution of charge. It is sometimes convenient when modeling bar magnets to assume that the end faces contain a distribution of fictitious magnetic charges or "poles." In terms of magnetic charges, we can express the surface and volume charge densities as [3]

$$\sigma_m = \vec{M} \cdot \hat{n}$$
$$\rho_m = -\nabla \cdot \vec{M} \tag{9.5}$$

and the scalar potential as

$$V_m = \frac{1}{4\pi} \int \frac{\sigma_m}{R} \, dS + \frac{1}{4\pi} \int \frac{\rho_m}{R} \, dV. \tag{9.6}$$

Thus in a bar magnet, we can assume that B comes from the Amperian currents flowing azimuthally around the cylinder and that H comes from magnetic charges on the flat end faces. In a real bar magnet, M is typically weaker near the end faces than it is in the central region. In this case, there will also be contributions to the field from the volume integrals above.

Consider a bar magnet with radius a that is much smaller than its length L. This is sometimes referred to as a "magnetic needle."[4] We look at the field on the axis outside the magnet at location z. We have

$$\cos \beta_2 = -\left\{1 + \left(\frac{a}{z - L/2}\right)^2\right\}^{-1/2}.$$

For $a \ll L$,

$$\cos \beta_2 \simeq -1 + \frac{1}{2}\left(\frac{a}{z - L/2}\right)^2$$

and we have a similar expression for $\cos \beta_1$ with $L \to -L$. Then from Equation 9.2,

$$B(z) = \frac{\mu_0 M \, a^2}{4(z - L/2)^2} - \frac{\mu_0 M \, a^2}{4(z + L/2)^2}.$$

If we define the strength of the magnetic charge as

$$q_m = \pi a^2 \, M, \tag{9.7}$$

we can write the magnetic intensity as

$$H(z) = \frac{q_m}{4\pi(z - L/2)^2} - \frac{q_m}{4\pi(z + L/2)^2}.$$

We interpret this as the field due to a positive magnetic charge at face 2 and a negative charge at face 1. The field strength falls off like the inverse square distance from the charge.

9.2 Magnetic circuit energized by a permanent magnet

Consider the arrangement shown in Figure 9.2, where a pair of permanent magnets surround an open gap on one side and are connected by an iron path on the other.[5] We know from the Ampère law that

$$\oint \vec{H} \cdot \vec{dl} = 0$$

around the circuit since there are no conduction currents. Assuming the gap is small and that the leakage flux is negligible, this gives

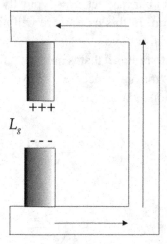

Figure 9.2 Magnetic circuit containing permanent magnets.

$$H_g L_g - H_m L_m + H_i L_i = 0,$$

where the subscripts g, m, and i refer to gap, magnet, and iron. The quantity $H_m L_m$ takes the place of NI in magnetic circuits powered by conductors. Neglecting leakage, the flux is constant around the loop, so we have

$$\Phi_B = B_i A_i = B_g A_g = B_m A_m,$$

where A is the cross-sectional area. Combining these equations, we have

$$H_m L_m = B_g A_g \left[\frac{H_g L_g}{B_g A_g} + \frac{H_i L_i}{B_g A_g} \right]$$

$$= B_m A_m \left[\frac{L_g}{\mu_0 A_g} + \frac{L_i}{\mu A_i} \right] \tag{9.8}$$

$$= \Phi_B R,$$

where the series reluctance is

$$R = \frac{L_g}{\mu_0 A_g} + \frac{L_i}{\mu A_i}.$$

Equation 9.8 is analogous to Ohm's law with $H_m L_m$ corresponding to the applied voltage.

The magnetic energy stored in the gap is

$$W_g = \tfrac{1}{2} \mu_0 B_g^2 A_g L_g$$

$$= \tfrac{1}{2} H_g L_g B_g A_g.$$

Since $\mu \gg \mu_0$, we can neglect the reluctance through the iron. Then

$$H_m L_m \simeq H_g L_g$$

and we can write

$$W_g \simeq \tfrac{1}{2} B_m H_m \, A_m L_m. \tag{9.9}$$

Thus the stored energy in the gap is proportional to the BH product and the volume of the permanent magnet.

Since B and H point in opposite directions, permanent magnets operate in the second quadrant of the hysteresis curve. To maximize the energy stored in the gap, it is desirable to operate at a point where the BH product is maximum. Rewriting Equation 9.8 as

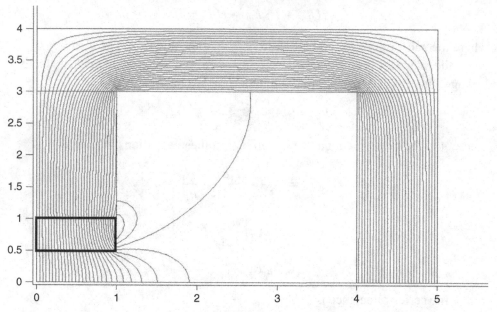

Figure 9.3 POISSON model of a magnetic circuit energized by a pair of permanent magnets.

$$\frac{B_m}{H_m} = \frac{L_m}{A_m R},$$

the quantities on the right-hand side can be adjusted to get the *B-H* "load line" for the magnetic circuit to pass through the point where the *BH* product is maximum.

Figure 9.3 shows a simple POISSON model[1] of a magnetic circuit energized by a pair of permanent magnets. The figure shows one quarter of a symmetric circuit. The *x* and *y* axes are symmetry planes in this figure. The use of a 2D program such as POISSON assumes that the configuration is uniform over a large distance in the third dimension (into the figure).

The permanent magnet material is located in the rectangle near the origin. The magnetization was oriented in the vertical direction. Considerable field fringing is evident in the air gap for this simple geometry. The amount of fringing depends strongly on the type of permanent magnet material that is used.[1]

9.3 Material properties

Characteristic properties of some permanent magnet materials are given in Table 9.1. Alnico is an alloy of iron with aluminum, nickel, and cobalt. It has

---

[1] The calculation was actually made with the PANDIRA program in the POISSON distribution.

Table 9.1 *Properties of permanent magnet materials [6, 7]*

	B_R **[T]**	H_C **[kOe]**[1]	H_{Ci} **[kOe]**	BH_{max} **[MG Oe]**
Alnico	0.83–1.25	0.6–6.4	0.6 – 1.9	1.4–5.5
Ceramic ferrite	0.22–0.39	1.9–3.2	2.5–3.3	1–3.6
SmCo	0.81–1.15	7.2–10.6	9–18	16–32
NdFeB	0.98–1.35	7.5–12.8	8–26	24–45

[1] 1 Oe = 1 10^{-4} T/ μ_0.

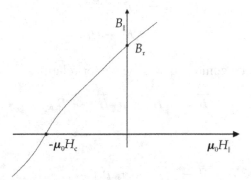

Figure 9.4 Second quadrant of the hysteresis curve.

a high remanent field but is easily demagnetized and has a large leakage flux.[1] Ferrite contains iron oxides (Fe_2O_3). It has a low B_R, but it is a cheap material. Ceramic ferrites are compounds of barium or strontium ferrite. Samarium and neodymium are rare earth elements.[8] They have large values of $(BH)_{max}$, can produce a large field from a compact magnet, are very resistant to demagnification, and have small leakage flux.

9.4 Model for rare earth materials

Usually there is some direction in a crystalline material along which the moments in the material tend to align. This is referred to as the "easy" magnetization direction. A permanent magnet can be produced so that the maximum magnetization is along some desired direction. A useful model has been developed for describing the magnetic properties of rare earth materials.[1, 9] Figure 9.4 shows the second quadrant of a typical hysteresis curve for the easy direction. It is a good approximation to assume the relative permeability in this region is equal to 1, so we have the relation

$$B_R \simeq \mu_0 H_C.$$

In this case, the fields from assemblies of blocks can be linearly superimposed. If the magnetization is anisotropic, the relation between B and H can be written as

$$B_\| = \mu_\| H_\| + B_R$$
$$B_\perp = \mu_\perp H_\perp, \tag{9.10}$$

where $\|$ refers to the easy direction in the material and $\perp$ refers to the direction perpendicular to it. For convenience, we define the *reluctivity* $\gamma = 1/\mu$. Then from Equation 9.10, we have

$$H_\| = \gamma_\| B_\| - \frac{B_R}{\mu_\|}$$
$$\simeq \gamma_\| B_\| - H_C. \tag{9.11}$$

These equations can be combined into the vector relations

$$\vec{B} = \mu_\| \vec{H}_\| + \mu_\perp \vec{H}_\perp + \vec{B}_R \tag{9.12}$$

and

$$\vec{H} = \gamma_\| \vec{B}_\| + \gamma_\perp \vec{B}_\perp - \vec{H}_C. \tag{9.13}$$

The vectors B_R and H_C are directed in opposite directions along the easy axis. Taking the divergence of Equation 9.12, we find that

$$\nabla \cdot (\mu_\| \vec{H}_\| + \mu_\perp \vec{H}_\perp) = -\nabla \cdot \vec{B}_R \equiv \rho_m, \tag{9.14}$$

which relates H to the density of magnetic charges. Taking the curl of Equation 9.13 in a region with no conduction currents, we get

$$\nabla \times (\gamma_\| \vec{B}_\| + \gamma_\perp \vec{B}_\perp) = \nabla \times \vec{H}_C \equiv \vec{J_m}, \tag{9.15}$$

which relates B to the Amperian currents. Thus the material can be treated magnetically as vacuum together with either a charge density $-\nabla \cdot \vec{B}_R$ or a current density $\nabla \times \vec{H}_C$. For homogeneous materials, these charges or currents vanish everywhere except on the surface.

The scalar potential for the permanent magnet material can be written as

$$\mu_0 V_m = \frac{1}{4\pi} \int \frac{\rho_m}{R} \, dV$$

$$= \frac{1}{4\pi} \int \frac{(-\nabla \cdot \vec{B}_R)}{R} \, dV, \tag{9.16}$$

where R is the distance between the source point and the field point. If the material is inhomogeneous, we can define $G = 1/R$ and make use of Equation B.3

$$\nabla \cdot (G\vec{B}_R) = G\nabla \cdot \vec{B}_R + \vec{B}_R \cdot \nabla G$$

to write Equation 9.16 in the form

$$\mu_0 V_m = -\frac{1}{4\pi} \int \left[\nabla \cdot (G\vec{B}_R) - \vec{B}_R \cdot \nabla G \right] dV.$$

Using the divergence theorem on the first part of the integrand,

$$\int \nabla \cdot (G\vec{B}_R) \, dV = \int G\vec{B}_R \cdot \hat{n} \, dS = 0,$$

which vanishes because we can choose the surface to lie outside the material where $B_R = 0$. We also have

$$\nabla G = -\frac{\vec{R}}{R^3}.$$

Thus the scalar potential for inhomogeneous material is

$$\mu_0 V_m = \frac{1}{4\pi} \int \frac{\vec{B}_R \cdot \vec{R}}{R^3} \, dV. \tag{9.17}$$

For homogeneous material, we can use the divergence theorem in Equation 9.16 and find that

$$\mu_0 V_m = -\frac{1}{4\pi} \int \frac{\vec{B}_R \cdot \hat{n}}{R} \, dS$$

$$= -\frac{\vec{B}_R}{4\pi} \cdot \int \frac{\hat{n}}{R} \, dS, \tag{9.18}$$

where the final integral only involves geometric quantities.

9.5 Rare earth model in two dimensions

If the permanent magnet pieces are uniform over a long distance in the z direction and the magnetization does not have a component along z, then it is possible to make a two-dimensional approximation of the fields.[9] From Equation 5.40, the field at the observation point z_o is

$$B^*(z_o) = \frac{\mu_0}{2\pi i} \iint \frac{\sigma}{z_o - z} \, dx \, dy. \qquad (9.19)$$

We assume the remanent field can have both x and y components

$$B_R = B_{Rx} + i \, B_{Ry}.$$

From the curl $B = \mu_0 \, J$ equation, we have

$$\partial_x B_{Ry} - \partial_y B_{Rx} = \mu_0 \, \sigma.$$

Substituting back into Equation 9.19, we get

$$B^*(z_o) = \frac{1}{2\pi i} [\mathbb{I}_1 - \mathbb{I}_2], \qquad (9.20)$$

where

$$\mathbb{I}_1 = \iint \frac{\partial_x B_{Ry}}{z_o - x - i \, y} \, dx \, dy$$

$$\mathbb{I}_2 = \iint \frac{\partial_y B_{Rx}}{z_o - x - i \, y} \, dx \, dy.$$

Considering the first integral, we integrate over x by parts with

$$u = \frac{1}{z_o - x - i \, y}$$

$$dv = \partial_x B_{Ry} \, dx,$$

which gives

$$\mathbb{I}_1 = \int \left[\frac{B_{R\,y}}{z_o - x - i \, y} - \int \frac{B_{R\,y}}{(z_o - x - i \, y)^2} \, dx \right] dy$$

$$= \oint \frac{B_{R\,y}}{z_o - x - i \, y} \, dy - \iint \frac{B_{R\,y}}{(z_o - x - i \, y)^2} \, dx \, dy.$$

The first term vanishes for a line integral evaluated outside the material where $B_R = 0$. Thus we have

$$\mathbb{I}_1 = -\iint \frac{B_{R\,y}}{(z_o - x - i \, y)^2} \, dx \, dy.$$

For the integral $\mathbb{I}_2$, we integrate over y by parts to find

$$\mathbb{I}_2 = -i \iint \frac{B_R \, x}{(z_o - x - i \, y)^2} \, dx \, dy.$$

Substituting back into Equation 9.20, we find that [9]

$$B^*(z_o) = \frac{1}{2\pi} \iint \frac{B_R}{(z_o - z)^2} \, dx \, dy. \tag{9.21}$$

If the material is homogeneous,

$$B^*(z_o) = \frac{B_R}{2\pi} \iint \frac{1}{(z_o - z)^2} \, dx \, dy$$

we can apply the complex Green's function, Equation 5.39, with

$$F = \frac{B_R}{2\pi} \frac{1}{z_o - z}.$$

The field for a homogeneous block can then be written in terms of the contour integral

$$B^*(z_o) = -\frac{B_R}{4\pi i} \oint \frac{dz^*}{z_o - z}. \tag{9.22}$$

It is also possible by direct integration to write this in the alternate forms [9]

$$B^*(z_o) = -\frac{B_R}{2\pi i} \oint \frac{dx}{z_o - z}$$

$$= \frac{B_R}{2\pi} \oint \frac{dy}{z_o - z}. \tag{9.23}$$

9.6 Multipole expansion for continuously distributed material

We next consider the problem of assembling permanent magnet material to make a 2D multipole magnet.[9, 10] In a multipole magnet, we try to make some multipole order N as large as possible, while at the same time making all the other orders small. We will mainly be concerned here with the magnetic field inside the aperture of the magnet. The field in general is given by Equation 9.21. In order to study the multipole structure of the field, it is convenient to first expand the denominator in a power series. We start with

$$\frac{1}{z_o - z} = \frac{-1}{z\left(1 - \frac{z_o}{z}\right)} = -\sum_{m=0}^{\infty} \frac{z_o^m}{z^{m+1}}.$$

If we let $n = m + 1$, then

$$\frac{1}{z_o - z} = -\sum_{n=1}^{\infty} \frac{z_o^{n-1}}{z^n}. \tag{9.24}$$

Taking the derivative with respect to z of both sides gives

$$\frac{1}{(z_o - z)^2} = \sum_{n=1}^{\infty} \frac{n\, z_o^{n-1}}{z^{n+1}}. \tag{9.25}$$

Substituting back into Equation 9.21, we have

$$B^*(z_o) = \frac{1}{2\pi} \sum_{n=1}^{\infty} n\, z_o^{n-1} \int \frac{B_R}{z^{n+1}}\, dS.$$

We identify the nth multipole field contribution as

$$B_n = \frac{n}{2\pi} \int \frac{B_R}{z^{n+1}}\, dS, \tag{9.26}$$

so the field in the aperture has the multipole expansion

$$B^*(z_o) = \sum_{n=1}^{\infty} B_n\, z_o^{n-1}. \tag{9.27}$$

It is possible to express B_n as a contour integral by defining

$$F = -\frac{B_R}{2\pi}\, z^{-n}$$

in the Green's theorem in Equation 5.39, which gives

$$B_n = \frac{B_R}{4\pi i} \oint \frac{dz^*}{z^n}. \tag{9.28}$$

Now let us consider the design of an ideal multipole magnet. We use a polar coordinate system with $z = r\, e^{i\phi}$. We assume the material is located in an annular region between the radii r_1 and r_2. Assume that we have the freedom to specify the direction of the easy axis everywhere in the material. Let the easy axis in the material located at angle ϕ be rotated through an angle $\beta(\phi)$ with respect to its direction at $\phi = 0$. The contribution to the field from the material at the angle ϕ is

$$B_r(\phi) = e^{i\beta(\phi)} B_R(0).$$

From Equation 9.26, the multipole field contribution is

$$B_n = \frac{n}{2\pi} \iint \frac{B_R \exp\{i[\beta(\phi) - (n+1)\phi]\}}{r^{n+1}} \, r \, d\phi \, dr. \tag{9.29}$$

The first term in the square brackets comes from the easy axis distribution and the second term comes from the azimuthal dependence in z^{n+1} in Equation 9.25. We can make the multipole as large as possible by making the quantity in square brackets equal to 0. This determines the constraint on the easy axis angles

$$\beta(\phi) = (n+1) \, \phi. \tag{9.30}$$

Multipole fields in the aperture can be produced by choosing n to be a positive integer N. For example, $N = 1$ produces a uniform dipole field in the aperture with no field outside the permanent magnet ring. Choosing n to be a negative integer on the other hand, produces a field outside the ring and no field in the aperture.[11] For example, $n = -1$ produces a dipole field outside the ring. Further control over the field can be obtained by surrounding the permanent magnet ring with an additional soft-iron ring.[11]

Confining our attention to $2N$-multipole fields inside the aperture, setting $n = N$ in Equation 9.30 and substituting into Equation 9.29 we find

$$B_n = \frac{n \, B_R}{2\pi} \int_{r_1}^{r_2} \frac{1}{r^n} \, \mathbb{I} \, dr,$$

where

$$\mathbb{I} = \int_0^{2\pi} e^{i \, (N-n)\phi} \, d\phi$$

$$= \begin{cases} 2\pi & \text{for } n = N \\ 0 & \text{for } n \neq N. \end{cases}$$

For a dipole assembly, $N = 1$, and we have

$$B_1 = B_R \ln\left(\frac{r_2}{r_1}\right), \tag{9.31}$$

while for higher order multipoles, $N \geq 2$, we get

$$B_N = N \, B_R \int_{r_1}^{r_2} \frac{dr}{r^N}$$

$$= \frac{N \, B_R}{N-1} \frac{1}{r_1^{N-1}} \left[1 - \left(\frac{r_1}{r_2} \right)^{N-1} \right]. \tag{9.32}$$

The $2N$-multipole field at the point z_o is given by

$$B^*(z_o) = B_N \, z_o^{N-1}.$$

9.7 Segmented multipole magnet assemblies

In practical terms, the continuous distribution of easy axis directions discussed in the previous section can only be approximated by using a finite number of magnetized blocks. As a result, the strength of the desired multipole is reduced and unwanted multipole orders are introduced.

Assume that the multipole magnet is constructed from M geometrically identical rare earth trapezoidal blocks.[9] Each block is separated from its nearest neighbor by the angle $2\pi/M$ around the z axis. In each block, the easy axis direction is chosen to approximate the ideal angle given in Equation 9.30. The direction of the easy axis inside a block is made to rotate by the angle $(N + 1) \, 2\pi/M$ from a given block to its following neighbor. For example, Figure 9.5 shows a dipole magnet ($N = 1$) made up of $M = 8$ rare earth blocks. In this assembly, neighboring blocks are

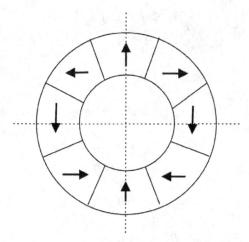

Figure 9.5 Dipole magnet made up from an assembly of trapezoidal-shaped permanent magnets.

separated geometrically by the angle $\pi/4$ and the easy axis rotates by $\pi/2$ from block to block.

Let C_n be the contribution to the multipole B_n for some reference block. Then the contribution of block m to B_n is

$$C_n \exp\left\{ i\, m\, \frac{2\pi}{M}\, (N+1) \right\} \exp\left\{ -i\, m\, \frac{2\pi}{M}\, (n+1) \right\},$$

where the first exponential comes from the rotation of the easy axis and the second exponential comes from the definition of B_n. The sum of all M blocks gives

$$B_n = C_n \sum_{m=0}^{M-1} \exp\left\{ i\, 2\pi m\, \frac{N-n}{M} \right\},$$

where $m = 0$ refers to the reference block. If the quantity $(N-n)/M$ is a positive or negative integer, then by writing the exponential in terms of cosines and sines, we see that the value of the sum is M. If $(N-n)/M$ is not an integer, then the value of the sum is zero.[9] Thus we have from Equation 9.27

$$B^*(z_o) = M \sum_{v=0}^{\infty} C_n\, z_o^{N-1}, \tag{9.33}$$

where v is an index to the set of allowed multipoles. The multipole order that corresponds to a given v is given by [9]

$$n = N + vM. \tag{9.34}$$

Next we turn to finding the reference multipole C_n. Substituting the series expansion Equation 9.24 into the contour integral 9.23, we find we can express C_n as

$$C_n = \frac{B_R}{2\pi i} \oint \frac{dx}{z^n}.$$

Figure 9.6 shows the geometry of a trapezoidal block. Only the two segments marked 1 and 2 contribute to the contour integral. On path 1 we have

$$y = -x \tan \alpha,$$

where $\alpha = \pi/M$. On path 2 we have

$$y = x \tan \alpha.$$

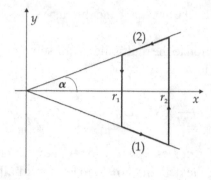

Figure 9.6 Trapezoidal block.

Thus the coefficient can be written as

$$C_n = \frac{B_R}{2\pi i} \left[\int_{r_1}^{r_2} \frac{dx}{(x - i x \tan \alpha)^n} + \int_{r_2}^{r_1} \frac{dx}{(x + i x \tan \alpha)^n} \right]$$

$$= \frac{B_R}{2\pi i} \left[\frac{1}{(1 - i \tan \alpha)^n} \int_{r_1}^{r_2} \frac{dx}{x^n} + \frac{1}{(1 + i \tan \alpha)^n} \int_{r_2}^{r_1} \frac{dx}{x^n} \right].$$

We can express

$$(1 \pm i \tan \alpha)^n = \frac{e^{\pm i n \alpha}}{\cos^n \alpha}.$$

Performing the integrals and rearranging terms, we get

$$C_n = -\frac{B_R \cos^n \alpha}{2\pi i (n - 1) r_1^{n-1}} \left[1 - \left(\frac{r_1}{r_2} \right)^{n-1} \right] (e^{-i n \alpha} - e^{i n \alpha}).$$

Writing the exponential term as a sine function, we find that C_n for the trapezoidal block is

$$C_n = \frac{B_R}{\pi (n - 1) r_1^{n-1}} \left[1 - \left(\frac{r_1}{r_2} \right)^{n-1} \right] \cos^n \alpha \sin n\alpha. \qquad (9.35)$$

Substituting this back into Equation 9.33, we find that the field inside the aperture is given by

$$B^*(z_o) = B_R \sum_{v=0}^{\infty} \left(\frac{z_o}{r_1} \right)^{n-1} \frac{n}{n - 1} \left[1 - \left(\frac{r_1}{r_2} \right)^{n-1} \right] K_n, \qquad (9.36)$$

Table 9.2 *Properties of a quadrupole magnet assembly*

M	K_2	$B_2(r_1)$ **[T]**	1st allowed harmonic [T]
4	0.32	0.454	$B_6(r_1) = -0.030$
8	0.77	1.095	$B_{10}(r_1) = -0.086$
12	0.89	1.270	$B_{14}(r_1) = -0.086$
16	0.94	1.336	$B_{18}(r_1) = -0.077$

where n is related to v by Equation 9.34 and we define the segmentation efficiency factor as [9]

$$K_n = \cos^n\left(\frac{\pi}{M}\right) \frac{\sin\left(\dfrac{n\pi}{M}\right)}{\dfrac{n\pi}{M}}. \tag{9.37}$$

The factor K_n measures how well a segmented magnet with M blocks approximates the idealized magnet with continuous variation of the easy axis that we described in the previous section.

Example 9.1: quadrupole magnet assembly

As an example, let us consider a quadrupole ($N = 2$) assembly made up of trapezoidal blocks. We assume $B_R = 0.95\ T$ and that $r_2/r_1 = 4$. We choose the observation point on the inner edge of the permanent magnet material, i.e., $z_o = r_1$. Table 9.2 shows some properties of the assembly as a function of the number of blocks M used around the circumference. The second column gives the segmentation efficiency factor for the quadrupole moment. We see that the efficiency increases quickly with the number of blocks used in the assembly. The third column gives the fundamental multipole contribution to the field. Already with 12 blocks, this contribution is almost 90% of the ideal unsegmented value. The last column shows the contribution to the field from the first allowed harmonic term. The strength of this term is ~7% of that for the fundamental.

Assemblies of permanent magnet blocks can be used to make many magnetic configurations, including dipoles, quadrupoles, sextupoles, solenoids, and periodic transverse fields.[1, 10]

References

[1] J. Bahrdt, *Permanent Magnets Including Undulators and Wigglers*, Proc. CERN Accelerator School on Magnets, Bruges, Belgium, June 2009, CERN-2010-004, p. 185.

[2] D. Tomboulian, *Electric and Magnetic Fields*, Harcourt, Brace & World, 1965, p. 240.

[3] L. Eyges, *The Classical Electromagnetic Field*, Dover, 1980, p. 139.

[4] W. Smythe, *Static and Dynamic Electricity*, 2nd ed., McGraw-Hill, 1950, p. 434.

[5] P. Lorrain & D. Corson, *Electromagnetic Fields and Waves*, 2nd ed., Freeman, 1970, p. 409.

[6] *High Performance Permanent Magnets*, Magnet Sales and Manufacturing, Inc., Culver City, CA, 1993.

[7] A. Chao & M. Tigner, *Handbook of Accelerator Physics and Engineering*, World Scientific, 1999, p. 366.

[8] J. Becker, Permanent magnets, *Sci. Am.* 223:92, 1970.

[9] K. Halbach, Design of permanent multipole magnets with oriented rare earth cobalt material, *Nuc. Instr. Meth.* 169:1, 1980.

[10] A. Chao & M. Tigner, *op. cit.*, p. 468.

[11] Q. Peng, S. McMurry & J. Coey, Cylindrical permanent magnet structures using images in an iron shield, *IEEE Trans. Mag.* 39:1983, 2003.

10

Time-varying fields

Until this point, we have only examined magnetic effects due to steady currents or magnetic materials in stationary configurations. In this chapter, we will partially relax this constraint by considering phenomena where there are slow variations in current or magnetic flux. By slow, we mean slow enough that we can ignore all effects of electromagnetic radiation. We begin with a discussion of Faraday's law, which presents another connection between electric and magnetic phenomena. This is followed by a more detailed discussion of the energy associated with a magnetic field, including the energy loss from the hysteresis cycle in ferromagnetic materials. We find that Faraday's law leads to the production of eddy currents in some materials, while the skin effect can restrict currents to a layer near the surface. We introduce the displacement current, which finally allows us to give a complete set of Maxwell's equations for stationary media. We conclude the chapter with a brief discussion of magnetic measurements.

10.1 Faraday's law

Michael Faraday discovered that a changing magnetic flux through a wire circuit C induced a voltage in the wire.

$$V \propto \frac{d\Phi_B}{dt} \tag{10.1}$$

The changing flux could be due to changing the current in C itself, changing the current in a second, nearby circuit, moving a second circuit or permanent magnet with respect to C, or changing the shape of C. Here we will mostly consider effects due to explicit changes in the current, in which case we can replace the total time derivative in Equation 10.1 with a partial derivative. According to *Lenz's law*, the voltage induced by the changing flux is such as to induce a current that gives rise to an additional flux that opposes the original change in flux. Thus Faraday's law can be written as

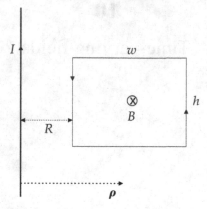

Figure 10.1 Electromotance induced in a rectangular loop.

$$\varepsilon = \oint \overrightarrow{E} \cdot \overrightarrow{dl} = -\frac{\partial}{\partial t} \int \overrightarrow{B} \cdot \hat{n} \, dS, \qquad (10.2)$$

where E is the *electric field intensity* and the surface S is bounded by the closed circuit. The field E acts on a distance element dl in its rest frame. Because of the tangential boundary condition on E, it follows that C can refer to any closed loop in space, not just a physical circuit.[1] The line integral[1] on the left side of Equation 10.2 is called the *electromotance* ε. If the flux links a coil with N turns, the electromotance must be multiplied by N. Since the contour integral is non-zero, the induced electric field in this case is nonconservative, i.e., work is done on a charge going around the contour.

Example 10.1: ε induced in a current loop

Consider a rectangular loop near a wire with increasing current I, shown in Figure 10.1. The time-dependent field from the wire is

$$\overrightarrow{B}(t) = \frac{\mu_0}{2\pi\rho} I(t) \, \hat{\phi}.$$

The flux through the square loop is

$$\Phi_B = \frac{\mu_0 I(t)}{2\pi} h \int_R^{R+w} \frac{d\rho}{\rho}$$

$$= \frac{\mu_0 I(t)}{2\pi} h \ln\left(\frac{R+w}{R}\right)$$

---

[1] Historically, ε has also been referred to as an *emf* or electromotive force.

and the electromotance is

$$\varepsilon = -\frac{\mu_0}{2\pi}\frac{dI}{dt}\,h\ln\left(\frac{R+w}{R}\right).$$

Using Stokes's theorem in Equation 10.2, we find

$$\int(\nabla\times\vec{E})\cdot\vec{dS} = -\frac{\partial}{\partial t}\int\vec{B}\cdot\vec{dS}$$

$$= -\int\frac{\partial\vec{B}}{\partial t}\cdot\vec{dS}.$$

Then since the surface S is arbitrary, we find the differential form of Faraday's law is

$$\nabla\times\vec{E} = -\frac{\partial\vec{B}}{\partial t}. \tag{10.3}$$

This equation is valid at any point in space. We can relate this to the vector potential by

$$\nabla\times\vec{E} = -\frac{\partial}{\partial t}(\nabla\times\vec{A})$$

$$= -\nabla\times\frac{\partial\vec{A}}{\partial t},$$

so that

$$\nabla\times\left(\vec{E}+\frac{\partial\vec{A}}{\partial t}\right) = 0.$$

Since its curl vanishes, the quantity in parentheses must be the gradient of a scalar function, which we denote V_e.

$$-\nabla V_e = \vec{E}+\frac{\partial\vec{A}}{\partial t}.$$

Thus the electric field

$$\vec{E} = -\nabla V_e - \frac{\partial\vec{A}}{\partial t}. \tag{10.4}$$

can arise from static charge distributions or from time-varying magnetic fields.

From Equation 1.32, the inductance is related to the flux by

$$LI = N\,\Phi_B.$$

Taking the time derivative of both sides, we find that an alternative definition of L is

$$L = -\frac{\varepsilon}{dI/dt}. \tag{10.5}$$

10.2 Energy in the magnetic field

We return now to the subject of the energy associated with a magnetic field. Consider a current element in an isolated loop together with an associated power source. Suppose that we want to increase the current in the loop from 0 up to some value I. For each step in the process of raising the current, the source must produce a voltage change

$$dV_e = -\nabla V_e \cdot \vec{dl}$$

across the current element and the source must supply the power

$$
\begin{aligned}
dP &= I\,dV_e \\
&= J\,dA\ (-\nabla V_e \cdot \vec{dl}),
\end{aligned}
$$

where A is the cross-sectional area of the conductor. Since J and dl are parallel, we can use Equation 10.4 to write

$$dP = \left(\vec{E} + \frac{\partial \vec{A}}{\partial t}\right) \cdot \vec{J}\ d\tau,$$

where, to minimize confusion, we use $d\tau$ in this section to represent the volume element. The total power provided by the source for the full loop is then [2]

$$P = \int \left(\vec{E} \cdot \vec{J} + \frac{\partial \vec{A}}{\partial t} \cdot \vec{J}\right) d\tau.$$

The first term on the right side is the power used to compensate for energy losses from heating in the conductor. The second term is the power used to set up the magnetic field associated with the increasing current, which is the subject of interest here. If we let W represent the energy stored in the magnetic field, then

$$\frac{dW}{dt} = \int \frac{\partial \vec{A}}{\partial t} \cdot \vec{J} \; d\tau. \tag{10.6}$$

Consider a small volume element of the conductor where J can be considered constant.[2] Then we can write A as the product of J with a factor that only depends on the geometry. Thus we can assume that $\vec{A} = \alpha \vec{J}$, where α is a constant. Then substituting

$$\frac{\partial}{\partial t} (\vec{A} \cdot \vec{J}) = 2\vec{J} \cdot \frac{\partial \vec{A}}{\partial t}$$

into Equation 10.6, we find the energy stored in the magnetic field is

$$W = \frac{1}{2} \int \vec{J} \cdot \vec{A} \; d\tau. \tag{10.7}$$

If the current distribution is a current loop, we let $\vec{J} \; d\tau \to I \; \vec{dl}$ and Equation 10.7 becomes

$$W = \frac{1}{2} I \int \vec{A} \cdot \vec{dl}.$$

This can be expressed in terms of the magnetic flux by

$$W = \frac{1}{2} I \; \Phi_B. \tag{10.8}$$

Returning again to Equation 10.7, we can use the curl $H = J$ equation to write the energy as

$$W = \frac{1}{2} \int (\nabla \times \vec{H}) \cdot \vec{A} \; d\tau.$$

Using the vector identity B.4, we find

$$W = \frac{1}{2} \int \left[\vec{H} \cdot (\nabla \times \vec{A}) - \nabla \cdot (\vec{A} \times \vec{H}) \right] d\tau.$$

Rewriting the first term in terms of B and using the divergence theorem in the second, we get

---

[2] J.D. Jackson, *Classical Electrodynamics*, Wiley, 1962, p. 176.

$$W = \frac{1}{2} \int \vec{H} \cdot \vec{B} \, d\tau + \frac{1}{2} \int \vec{H} \times \vec{A} \cdot \hat{n} \, dS.$$

Looking at the surface integral, we know that the field from a conductor element falls off like $1/R^2$ and the vector potential falls off like $1/R$, while the surface area only grows like R^2. By evaluating at a sufficiently large distance, the second integral vanishes. Thus the energy stored in the magnetic field from conduction currents is

$$W = \frac{1}{2} \int \vec{B} \cdot \vec{H} \, d\tau \qquad (10.9)$$

and the magnetic energy density in the field is

$$w_B = \frac{1}{2} \, \vec{B} \cdot \vec{H}. \qquad (10.10)$$

The energy of a permeable body with magnetization M in an applied magnetic field B_a can be expressed as [3]

$$W = \frac{1}{2} \int \vec{M} \cdot \vec{B} \, d\tau. \qquad (10.11)$$

10.3 Energy loss in hysteresis cycles

Consider a Rowland ring containing a ferromagnetic sample, as discussed in Section 2.5. If we increase the current in a conductor wound around the sample, we get an induced electromotance that opposes the change in current. The extra power expended by the source is

$$\frac{dW}{dt} = NI \, \frac{d\Phi_B}{dt}$$

$$= NIA \, \frac{dB}{dt}$$

$$= \frac{NI}{l} \, Al \, \frac{dB}{dt},$$

where N is the number of conductor turns, A is the cross-sectional area of the sample, l is the mean circumference of the ring, and B is the average flux density inside the sample. Using the Ampère law, this can be written

$$\frac{dW}{dt} = HV \, \frac{dB}{dt},$$

where V is the volume of the sample.

Now consider the hysteresis loop shown in Figure 10.2. The energy supplied by the source in moving from point a to point b along the loop is

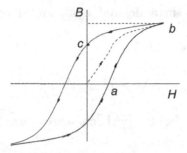

Figure 10.2 Energy loss in a hysteresis loop.

$$W_{ab} = V \int_a^b H \, dB.$$

Since dB is the independent variable, the value of this integral is the area projected on the B (vertical) axis in the figure. Going from point b to point c along the loop, I is in the same direction, but is decreasing. Thus the electromotance changes sign and some energy is returned to the source.

$$W_{bc} = -V \int_b^c H \, dB.$$

The sum of these two integrals is the area inside the hysteresis loop in the first quadrant. If we continue this analysis for a complete cycle, we find that the net energy lost in the ferromagnetic material per cycle is [4]

$$W = V \oint H \, dB. \tag{10.12}$$

This energy loss can be minimized by choosing ferromagnetic materials with a narrow hysteresis loop.

10.4 Eddy currents

Faraday's law shows that time-varying magnetic fields produce a voltage in materials such as conductors, iron, or mechanical supports. If a closed path exists inside the material, this voltage can drive currents, known as *eddy currents*.[5] The eddy currents can in turn create new magnetic fields that are superimposed over the original field. Eddy currents can be used for a number of desirable purposes, including displacement and position measurements, induction heating, magnetic shielding, levitation, and braking. On the other hand, undesirable effects from eddy currents include resistive power losses, Lorentz forces, multipole errors in a desired field, and a time lag in reaching an equilibrium field value.

Starting from Faraday's law

$$\nabla \times \vec{E} = -\frac{\partial \vec{B}}{\partial t},$$

multiplying both sides by the electrical conductivity σ and taking the curl, we find

$$\nabla \times (\nabla \times \sigma \vec{E}) = -\sigma \frac{\partial}{\partial t}(\nabla \times \vec{B}).$$

We can write Ohm's law in the form

$$\vec{J} = \sigma \vec{E}. \tag{10.13}$$

The range of current densities over which this linear relation holds depends on the material. Thus we have

$$\nabla \times (\nabla \times \vec{J}_e) = -\sigma \mu \frac{\partial}{\partial t}(\nabla \times \vec{H}),$$

where J_e is the eddy current density. We can use the vector identity B.7 on the left side of this equation and the curl $H = J$ equation on the right side to get

$$\nabla(\nabla \cdot \vec{J}_e) - \nabla^2 \vec{J}_e = -\sigma \mu \frac{\partial \vec{J}_e}{\partial t}.$$

Since the divergence term on the left side vanishes, we find that [6]

$$\nabla^2 \vec{J}_e = \sigma \mu \frac{\partial \vec{J}_e}{\partial t}. \tag{10.14}$$

This is a form of the diffusion equation. The rate of build-up of the eddy currents is controlled by the factor $\sigma \mu$.

If instead, we begin by taking the curl of the Ampère law, we find

$$\nabla \times (\nabla \times \vec{H}) = \nabla \times \vec{J}_e$$
$$= \sigma \nabla \times \vec{E}.$$

Applying Equation B.7 on the left-hand side and Faraday's law on the right, we obtain the equation

$$\nabla^2 \vec{H} = \sigma \mu \frac{\partial \vec{H}}{\partial t}. \tag{10.15}$$

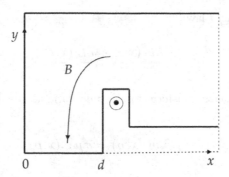

Figure 10.3 One-quarter of a symmetric *H*-dipole.

Thus the magnetic field associated with the eddy currents also satisfies a diffusion equation with the same characteristic constant. If one specifies the time dependence for *H* and the geometry of the configuration, the diffusion equation can be solved for the spatial and time dependence of the magnetic field due to eddy currents.[5] This may lead to a series of terms, each with its own characteristic time dependences.

Example 10.2: time constant for eddy currents in a solid iron core
Consider a long *H*-dipole with a solid iron yoke, as shown in Figure 10.3. For slow time changes, eddy currents can flow throughout the volume of the iron yoke surrounding the coil.[7] The magnetic flux from the eddy currents is not symmetric with the flux from the coils, which causes the iron saturation to vary with transverse position.

Consider a path through the iron yoke at the midplane in the region $0 \leq x \leq d$. Assuming there is no leakage flux, all of the return flux from the conductor has to pass across this path. Assume the current in the conductor is changing with time. Then the magnetic field in the vicinity of the path is in the *y* direction, the induced electric field is in the *z* direction, and on the midplane both fields are only functions of *x* and *t*.

$$\vec{B} = B_y(x,t)\,\hat{y}$$
$$\vec{E} = E_z(x,t)\,\hat{z}.$$

The eddy currents flow parallel to the *z* axis until they reach the magnet end, where they reverse direction and flow back at the symmetric (*x*, *y*) position on the other side of the magnet. From the Ampère and Ohm's laws, we have

$$\frac{\partial B_y(x,t)}{\partial x} = \sigma \mu \, E_z(x,t),$$

while from Faraday's law

$$\frac{\partial B_y(x,t)}{\partial t} = \frac{\partial E_z(x,t)}{\partial x}.$$

Applying the Laplace transform to the time variable for these two equations,[8] we get

$$\partial_x \mathcal{B}_y(x,p) = \sigma \mu \, \varepsilon_z(x,p) \tag{10.16}$$

and

$$p \, \mathcal{B}_y(x,p) = \partial_x \, \varepsilon_z(x,p), \tag{10.17}$$

where p is the variable conjugate to t in the Laplace transform. Taking the derivative of Equation 10.16 with respect to x and substituting Equation 10.17, we get

$$\partial_x^2 \mathcal{B}_y(x,p) = \sigma \mu p \, \mathcal{B}_y(x,p).$$

Defining $k^2 = \sigma \mu p$, the solution for the magnetic field consistent with the boundary conditions is [7]

$$\mathcal{B}_y(x,p) = \mathcal{B}_0 \cosh kx.$$

The field across the return yoke is asymmetric and is larger on the side nearer the coil. At the edge of the path closest to the conductor, we have

$$\begin{aligned} kd &= d\sqrt{\sigma \mu p} \\ &= \sqrt{\sigma \mu p \, d^2} \\ &= \sqrt{p\,\tau}, \end{aligned}$$

where [7]

$$\tau = \sigma \mu d^2. \tag{10.18}$$

The variable τ has the dimensions of time. It gives a characteristic time for eddy current effects in this configuration. Note that it depends quadratically on the width d of the return yoke.

Eddy currents can be suppressed by restricting the rate of change of the desired field or by constructing the magnet in such a way that potential eddy current loops are minimized. Magnet yokes are frequently constructed by assembling thin iron laminations for this reason.

10.5 Skin effect

Consider a current density with the periodic time variation

$$\vec{J} = \vec{J}_0\, e^{i\omega t},$$

where ω is the angular frequency and J_0 only depends on the spatial dimensions. The diffusion equation for the current density, analogous to Equation 10.14, is

$$\nabla^2 \vec{J} = \sigma\mu \frac{\partial \vec{J}}{\partial t}$$

$$= i\,\omega\sigma\mu\, \vec{J}.$$

Defining $\zeta^2 = i\omega\sigma\mu$, we obtain

$$\nabla^2 \vec{J} - \zeta^2 \vec{J} = 0. \tag{10.19}$$

Now assume that the current is flowing along a conducting slab that occupies the space $y \le 0$. Then the component of J flowing in the z direction, for example, is

$$J_z = J_{z0}\, e^{i\omega t}.$$

Applying Equation 10.19 to this, we find

$$\frac{d^2 J_{z0}}{dy^2} = \zeta^2 J_{z0}.$$

This differential equation has the solution

$$J_{z0} = J_S\, e^{-\zeta |y|},$$

where J_S is the spatial dependence of the current density on the surface of the slab. Using the relation

$$i = \tfrac{1}{2}\,(1 + 2\,i - 1) = \tfrac{1}{2}\,(1 + i)^2,$$

we can write

$$\zeta = \pm \frac{1}{\sqrt{2}}\,(1 + i)\,\sqrt{\omega\sigma\mu}.$$

The boundary condition for large $|\,y\,|$ eliminates the negative solution for ζ. Thus

$$\zeta = (1 + i)\,\sqrt{\frac{\omega\sigma\mu}{2}}.$$

Defining the *skin depth* as

$$\delta = \sqrt{\frac{2}{\omega\sigma\mu}},$$ (10.20)

the solution for the current density is [9]

$$J_z = J_S \, \exp\left\{-\frac{|y|}{\delta}\right\} \exp\left\{i\left(\omega t - \frac{|y|}{\delta}\right)\right\}.$$ (10.21)

We see that the current density decreases exponentially with distance into the surface. In addition, there is a phase shift of the current flowing inside the material with respect to the current flowing on the surface. These effects scale with the skin depth parameter δ. For a copper conductor with current varying at 1 kHz, the skin depth is ~2.1 mm.

10.6 Displacement current

We have seen in Chapter 1 that $\nabla \cdot \vec{J} = 0$ in magnetostatic problems. However, once we allow for time variations, the conservation of charge requires

$$\nabla \cdot \vec{J} + \frac{\partial \rho}{\partial t} = 0,$$ (10.22)

where ρ is the electric charge density of free (i.e., unbound) charges. Therefore, when time variation is allowed, the divergence of the conduction current density no longer needs to vanish. From electrostatics, we know that [10]

$$\nabla \cdot \vec{D} = \rho,$$ (10.23)

The vector D is called the *electric flux density*[3] and is related to the electric field intensity by

$$\vec{D} = \varepsilon \, \vec{E}$$ (10.24)

for linear materials, where ε is the *permittivity*. Taking the time derivative of Equation 10.23, we get

$$\frac{\partial \rho}{\partial t} = \nabla \cdot \frac{\partial \vec{D}}{\partial t}.$$ (10.25)

---

[3] Historically, the vector D is also known as the electric displacement.

Table 10.1 *Maxwell's equations*

Differential form	Integral form
$\nabla \cdot \vec{D} = \rho$	$\int \vec{D} \cdot \vec{dS} = \int \rho \, dV$
$\nabla \times \vec{E} = -\dfrac{\partial \vec{B}}{\partial t}$	$\oint \vec{E} \cdot \vec{dl} = -\int \dfrac{\partial \vec{B}}{\partial t} \cdot \vec{dS}$
$\nabla \cdot \vec{B} = 0$	$\int \vec{B} \cdot \vec{dS} = 0$
$\nabla \times \vec{H} = \vec{J} + \dfrac{\partial \vec{D}}{\partial t}$	$\oint \vec{H} \cdot \vec{dl} = \int \left(\vec{J} + \dfrac{\partial \vec{D}}{\partial t} \right) \cdot \vec{dS}$

Comparing Equations 10.22 and 10.25, we see that the quantity $\partial \vec{D}/\partial t$ acts like an additional kind of current. Thus we define the displacement current density as

$$\vec{J}_d = \frac{\partial \vec{D}}{\partial t}. \tag{10.26}$$

Taking this into account, the Ampère law must then be modified as [11]

$$\nabla \times \vec{H} = \vec{J} + \frac{\partial \vec{D}}{\partial t}. \tag{10.27}$$

This shows that a magnetic field can also be produced by a time-varying electric field.

At this point, we can summarize the complete set of Maxwell's equations for stationary media in Table 10.1. It is important to keep in mind that writing the equations in this form assumes the validity of the *constitutive relations*

$$\vec{B} = \mu \vec{H}$$
$$\vec{J} = \sigma \vec{E}$$
$$\vec{D} = \varepsilon \vec{E}.$$

10.7 Rotating coil measurements

Magnetic fields can be measured using a number of techniques. Nuclear magnetic resonance (NMR) probes are used for high-precision measurements.[12] Hall effect probes are simple to use and are commercially available.[13] Other common methods of measuring the magnetic field are based on electromagnetic induction.

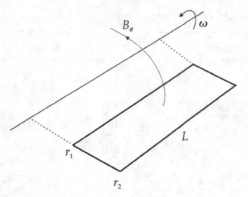

Figure 10.4 Field measurement with a radial coil.

One technique, which relates directly with our previous discussions of the multipole content of fields, involves measurements in long magnets with large aperture using a rotating coil.[14] The azimuthal component of the field can be measured using a radial coil, the principle of which is shown in Figure 10.4. The flux through the wire loop with N turns is

$$\Phi_B(\theta) = N L \int_{r_1}^{r_2} B_\theta(r, \theta)\, dr$$

$$= N L \sum_{n=1}^{\infty} (A_n \sin n\theta + B_n \cos n\theta) \int_{r_1}^{r_2} r^{n-1}\, dr$$

$$= N L \sum_{n=1}^{\infty} (A_n \sin n\theta + B_n \cos n\theta) \left[\frac{r_2^n - r_1^n}{n} \right],$$

where we have used Equation 4.8 to express the azimuthal field in terms of multipole field components. If the coil rotates at a constant rate, we have $\theta = \omega t$ and

$$\frac{d\Phi_B}{dt} = \frac{d\Phi_B}{d\theta} \frac{d\theta}{dt} = \omega \frac{d\Phi_B}{d\theta}.$$

The induced voltage in the coil from Faraday's law is then

$$V(\theta) = -\omega N L \sum_{n=1}^{\infty} (A_n \cos n\theta - B_n \sin n\theta)\, (r_2^n - r_1^n). \qquad (10.28)$$

Performing a Fourier analysis on the voltage signal allows the multipole coefficients to be determined.

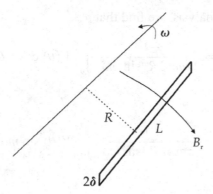

Figure 10.5 Field measurement with a tangential coil.

$$\int_0^{2\pi} V(\theta) \sin m\theta \, d\theta = -\omega NL \sum_{n=1}^{\infty} (r_2^n - r_1^n) \left[A_n \int_0^{2\pi} \cos n\theta \, \sin m\theta \, d\theta \right.$$

$$\left. -B_n \int_0^{2\pi} \sin n\theta \, \sin m\theta \, d\theta \right] = \omega NL(r_2^n - r_1^n) \, B_n \, \pi \, \delta_{mn}.$$

Thus

$$B_m = \frac{1}{\pi \, \omega N \, L \, (r_2^m - r_1^m)} \int_0^{2\pi} V(\theta) \sin m\theta \, d\theta. \tag{10.29}$$

Similarly we find that

$$A_m = \frac{-1}{\pi \, \omega N \, L \, (r_2^m - r_1^m)} \int_0^{2\pi} V(\theta) \cos m\theta \, d\theta. \tag{10.30}$$

It is possible to do a similar analysis on the radial component of the field B_r using the rotating tangential coil illustrated in Figure 10.5. Using Equation 4.7, we have

$$\Phi_B(\theta) = NL \int_{\theta-\delta}^{\theta+\delta} B_r(R, \theta) \, R \, d\theta$$

$$= 2NL \sum_{n=1}^{\infty} \frac{R^n}{n} \sin n\delta \, (-A_n \cos n\theta + B_n \sin n\theta).$$

The induced voltage in this case is

$$V(\theta) = -2\omega N \, L \sum_{n=1}^{\infty} R^n \sin n\delta \, (A_n \sin n\theta + B_n \cos n\theta). \tag{10.31}$$

Performing a Fourier analysis, we find that

$$B_m = \frac{-1}{2\pi\omega NLR^m \sin m\delta} \int_0^{2\pi} V(\theta) \cos m\theta \, d\theta \qquad (10.32)$$

and

$$A_m = \frac{-1}{2\pi\omega NLR^m \sin m\delta} \int_0^{2\pi} V(\theta) \sin m\theta \, d\theta. \qquad (10.33)$$

References

[1] W. Panofsky & M. Phillips, *Classical Electricity and Magnetism*, 2nd ed., Addison-Wesley, 1962, p. 159.

[2] P. Lorrain & D. Corson, *Electromagnetic Fields and Waves*, 2nd ed., Freeman, 1970, p. 351–356.

[3] J. Stratton, *Electromagnetic Theory*, McGraw-Hill, 1941, p. 126–128.

[4] P. Lorrain & D. Corson, *op. cit.*, p. 398–400.

[5] G. Moritz, Eddy currents in accelerator magnets, in D. Brandt (ed.), *Proc. CERN Accelerator School on Magnets*, Bruges, Belgium, June 2009, CERN-2010-004, p. 103.

[6] G. Harnwell, *Principles of Electricity and Magnetism*, 2nd ed., McGraw-Hill, 1949, p. 340–341.

[7] K. Halbach, Some eddy current effects in solid core magnets, *Nuc Instr. Meth.* 107:529, 1973.

[8] E. Kreyszig, *Advanced Engineering Mathematics*, Wiley, 1962, chapter 4.

[9] J. Marion, *Classical Electromagnetic Radiation*, Academic Press, 1965, p. 149.

[10] P. Lorrain & D. Corson, *op. cit.*, p. 105–106.

[11] J. D. Jackson, *Classical Electrodynamics*, Wiley, 1962, p. 177–178.

[12] C. Reymond, Magnetic resonance techniques, in S. Turner (ed.), *CERN Accelerator School on Measurement and Alignment of Accelerator and Detector Magnets*, CERN 98-05, 1998, p. 219.

[13] J. Kvitkovic, Hall generators, in S. Turner (ed.), ibid., p. 233.

[14] A. Jain, Harmonic coils, in S. Turner (ed.), ibid., p. 175.

11

Numerical methods

Until now, we have mostly considered magnetostatic problems that have analytic solutions. In practice, this has usually restricted our choice of problems to situations where one of the problem boundaries coincides with a coordinate system axis and to solutions that can be written as products of functions of a single coordinate. Frequently, more complicated problems can only be solved using numerical methods.[1] A number of commercial and freeware programs are available for solving magnetic problems. A lot of effort has been devoted to making many of these programs accurate, efficient, and user-friendly. If such a program is available and can address the problem under consideration, it is often the best choice to use it. However, there are occasions when new code must be written to solve a problem. It is also important to have some basic understanding about the methods involved in obtaining these solutions. In this chapter, we will examine three numerical methods that have been used for solving boundary value problems involving the Poisson equation: finite differences, finite elements, and integral equations. In each case, the analytical equation or its solution is approximated in some way that leads to a matrix equation for the unknown potential or field. We conclude the chapter with a discussion of inverse problems and optimization techniques.

11.1 Finite difference method

In the finite difference method, the continuous space of the problem domain is replaced with a grid of discrete points called *nodes*.[1] The grid, which is commonly rectangular or polar, must extend over the whole space of the problem. This usually requires grid points outside all conductors and iron, out to a point where the potential has some assumed value, typically zero. Symmetries in the configuration may be used to reduce the required grid size. For some problems, it may be necessary to use reduced grid spacing in regions where the field gradient is large or where high accuracy is required.

Unknown quantities are calculated at the nodes. Derivatives defined on the continuous domain of the physical problem are replaced with difference approximations defined in terms of the values of the unknown function at the nodes. For example, assume that u is an unknown function and that a one-dimensional problem space has been discretized with the grid spacing $h = \Delta x$. The Taylor series for the node at location $x + h$ is

$$u(x+h) = u(x) + h\frac{\partial u}{\partial x}\bigg|_x + \frac{h^2}{2!}\frac{\partial^2 u}{\partial x^2}\bigg|_x + \mathcal{O}(h^3). \tag{11.1}$$

Ignoring second- and higher-order terms, the first derivative can be approximated in terms of the node values by

$$\frac{\partial u}{\partial x} \simeq \frac{u(x+h) - u(x)}{h}. \tag{11.2}$$

This is called the *forward difference* because it involves the next higher node than the node at x. Similarly taking $h \rightarrow -h$, we find the *backward difference* is

$$\frac{\partial u}{\partial x} \simeq \frac{u(x) - u(x-h)}{h}. \tag{11.3}$$

We can obtain an approximation for the derivative that is accurate through the h^2 term in the Taylor series by calculating

$$u(x+h) - u(x-h) \simeq u + hu' + \frac{h^2}{2}u'' - u + hu' - \frac{h^2}{2}u''.$$

The *central difference* approximation for the first derivative is then

$$\frac{\partial u}{\partial x} \simeq \frac{u(x+h) - u(x-h)}{2h}. \tag{11.4}$$

We can approximate the second derivative from its definition as

$$\frac{\partial^2 u}{\partial x^2} \simeq \frac{1}{h}\left[\frac{u(x+h) - u(x)}{h} - \frac{u(x) - u(x-h)}{h}\right]$$

$$= \frac{1}{h^2}[u(x+h) - 2u(x) + u(x-h)]. \tag{11.5}$$

Example 11.1: one-dimensional Poisson equation
To illustrate the basic concepts of the finite difference method, let us consider the solution of the one-dimensional Poisson's equation

$$\frac{\partial^2 u}{\partial x^2} = f$$

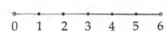

Figure 11.1 Nodes on a line.

in the greatly simplified situation shown in Figure 11.1. The line is discretized into 7 nodes. Let u refer to the unknown quantity, which we assume satisfies the Dirichlet boundary conditions $u_0 = 0$ and $u_6 = 0$. Assume that the source function f has the value $f_3 = v$ at the center of the line and is 0 otherwise. The values of u at the five interior nodes are the unknown quantities. Using Equation 11.5, each interior node satisfies the equation

$$u(x + h) - 2u(x) + u(x - h) = h^2 f.$$

We can write the equations for the five unknowns in the form of a matrix equation

$$Cu = g. \tag{11.6}$$

For the case here, we have

$$\begin{bmatrix} -2 & 1 & 0 & 0 & 0 \\ 1 & -2 & 1 & 0 & 0 \\ 0 & 1 & -2 & 1 & 0 \\ 0 & 0 & 1 & -2 & 1 \\ 0 & 0 & 0 & 1 & -2 \end{bmatrix} \begin{bmatrix} u_1 \\ u_2 \\ u_3 \\ u_4 \\ u_5 \end{bmatrix} = \begin{bmatrix} 0 \\ 0 \\ v \\ 0 \\ 0 \end{bmatrix},$$

which has the solution

$$u = \begin{bmatrix} -\tfrac{1}{2} \, h^2 v \\ -h^2 v \\ -3/2 \, h^2 v \\ -h^2 v \\ -\tfrac{1}{2} \, h^2 v \end{bmatrix}.$$

For a square grid in two dimensions, let us designate the node under consideration as node 0 and its four nearest neighbors as nodes 1–4, as shown in Figure 11.2. We can write the Laplacian operator in terms of the values at the five nodes as [2]

$$\nabla^2 u \simeq \frac{1}{h^2} [u(x - h, y) + u(x + h, y) + u(x, y - h) + u(x, y + h) - 4u(x, y)]. \tag{11.7}$$

It is also possible to write a generalized version for the two-dimensional Laplacian where the distances from a given node to each of its nearest neighbors can be different. If h is the characteristic grid spacing, then [3]

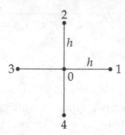

Figure 11.2 Node structure in two dimensions.

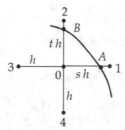

Figure 11.3 Node near a curved boundary.

$$\nabla^2 u \simeq \frac{2}{h^2}\left[\frac{u_1}{p(p+r)} + \frac{u_2}{q(q+s)} + \frac{u_3}{r(p+r)} + \frac{u_4}{s(q+s)} - \left(\frac{1}{pr} + \frac{1}{qs}\right)u_0\right], \quad (11.8)$$

where p, q, r, s are dimensionless scaling factors for the spacings from node 0 to its nearest neighbors. Using this expression for problems where the physical boundaries of conductors and iron are parallel to the x and y axes, it is possible to set up the equations for the interior nodes together with nodes coinciding with the boundaries. Higher-order difference equations for the Laplacian are also possible.[4]

A complication arises in setting up the node equations for nodes adjacent to boundaries that do not align exactly with the grid spacing, for example nodes next to circular boundaries in a rectangular grid. For Dirichlet boundary conditions, we can make use of the fact that we know the value of $u(x,y)$ on the boundary. Consider a node 0 adjacent to the boundary shown in Figure 11.3. The two-dimensional Laplacian operator acting at u_0 can be approximated as [5]

$$\nabla^2 u \simeq \frac{2}{h^2}\left[\frac{u_A}{s(1+s)} + \frac{u_B}{t(1+t)} + \frac{u_3}{1+s} + \frac{u_4}{1+t} - \left(\frac{1}{s} + \frac{1}{t}\right)u_0\right], \quad (11.9)$$

where s and t are dimensionless scale factors.

There are also complications in setting up the difference equations when the problem requires Neumann boundary conditions.[6] Here we will only consider the situation shown in Figure 11.4, which is a planar boundary in a square grid between

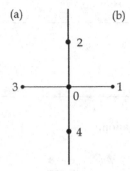

Figure 11.4 Boundary between regions with different permeabilities.

two regions (a) and (b) with different permeabilities. Let us consider an arbitrary point 0 along the boundary. Since point 0 is part of region a, the Laplace equation is

$$A_{a1} + A_2 + A_{a3} + A_4 - 4A_0 = 0, \tag{11.10}$$

where we use A for the unknown function here. For region a, node 1 is fictitious and must be eliminated from the final difference equation. Node 0 is also a part of region b, so we have

$$A_{b1} + A_2 + A_{b3} + A_4 - 4A_0 = 0, \tag{11.11}$$

where node 3 is fictitious in region b. The Neumann boundary condition at node 0 is

$$\frac{1}{\mu_a}\left(\frac{A_{a1} - A_{a3}}{2h}\right) = \frac{1}{\mu_b}\left(\frac{A_{b1} - A_{b3}}{2h}\right).$$

Substituting for A_{a1} from Equation 11.10 and A_{b3} from Equation 11.11, we find the difference equation at boundary point 0 is

$$4A_0 - \frac{2\mu_a}{\mu_a + \mu_b}A_{b1} - A_2 - \frac{2\mu_b}{\mu_a + \mu_b}A_{a3} - A_4 = 0. \tag{11.12}$$

If region b has infinite permeability, then the difference equation simplifies to

$$4A_0 - A_2 - 2A_{a3} - A_4 = 0. \tag{11.13}$$

Interpolation must be used when a value of some quantity u is required at a location away from one of the nodes. Suppose we want the value of the function $u(x, y)$, as shown for a rectangular grid in Figure 11.5. The simplest scheme for estimating the value of u is *bilinear interpolation*. We first determine which rectangle in the grid that the desired point is located in. Then defining the variables

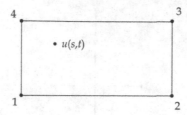

Figure 11.5 Bilinear interpolation.

$$s = \frac{x - x_1}{x_2 - x_1}$$

$$t = \frac{y - y_1}{y_2 - y_1},$$

we can approximate the value of $u(x, y)$ as [7]

$$u(s, t) \simeq (1 - s)(1 - t)u_1 + s(1 - t)u_2 + stu_3 + (1 - s)tu_4. \qquad (11.14)$$

This expression varies continuously in x and y and reduces correctly to the node values at the corners of the rectangle.

The discrepancy between the result from using the difference approximation and the exact result from solving the differential equation is known as the truncation error. The error can be estimated by examining the first term in the Taylor series that was neglected in deriving the difference formula under consideration. For a square mesh, the error on the second derivative goes like

$$\sim \frac{2h^2}{4!} \frac{\partial^4 u}{\partial x^4}\Big|_0.$$

The error is proportional to h^2, so one method of improving the accuracy in a finite difference calculation is to reduce the mesh spacing. We can monitor the improvement in accuracy by finding the maximum absolute value for the difference

$$e_{ij} = u_{ij}^{h_2} - u_{ij}^{h_1},$$

where the superscript refers to the mesh spacing used for the solution and (i, j) refers to nodes common to both mesh spacings. This approach is ultimately limited by the growth in the size of the coefficient matrix and by rounding errors in the numerical calculations. An alternative approach for increasing the accuracy of the calculation is to use higher-order difference equations.

The quality of a solution can be monitored by calculating the *residual* for each of the interior nodes. For a general node for the Poisson equation, the residual is defined as

$$R_{i,j} = 4 \ u_{i,j} - u_{i,j+1} - u_{i,j-1} - u_{i+1,j} - u_{i-1,j} - h^2 \mu_{i,j} J_{i,j}. \quad (11.15)$$

If the difference equation exactly satisfies Poisson's equation, the residual should be 0.

For problems using iteration algorithms, we can compute

$$e_{ij} = u_{ij}^n - u_{ij}^{n-1}$$

for the unknown function at the node (i,j), where the superscript refers to the iteration number. For these methods, one can estimate the quality of the solution by calculating the difference at all the nodes. Let M refer to the absolute value of the largest difference in the mesh.

$$M = \max \ |e_{ij}|$$

For the 5-point Laplacian operator in Equation 11.7, the error ε between the exact solution of the difference equation and the approximate solution after n iterations is bounded by [8]

$$\varepsilon \le \frac{M\rho^2}{4h^2}, \quad (11.16)$$

where ρ is the radius of the smallest circle that encompasses the entire field region.

In problems where iron saturation is a consideration, the permeability of the iron can be made a variable at each of the nodes in the iron regions.[9] The permeabilities are stored on a separate mesh. After each iteration of the potential, the field in the iron region is updated. A table of B-μ values can be used to relate the permeability to the field at the node. The mesh of permeability values is then updated using, for example, an under-relaxation algorithm.

11.2 Example solution using finite differences

As a simple example, let us consider a rectangular conductor with constant current density J close to a sheet of iron with permeability $\mu_r = 100$, as shown in Figure 11.6.

Assume that the conductor and the sheet are uniform in the z direction, so that a two-dimensional analysis is justified. Assume that the figure is up-down symmetric, so that the x axis lies in a symmetry plane. We solve the problem using finite differences on a square 200×200 mesh. For simplicity, we have chosen the boundaries of the conductor and the iron sheet to line up with node locations. The problem requires Dirichlet boundary conditions on the left, right, and top outer borders, where we set $A_z = 0$. Because of the up-down symmetry, the bottom border requires a Neumann boundary condition.

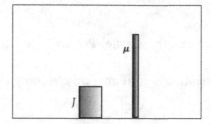

Figure 11.6 Conductor close to an iron sheet.

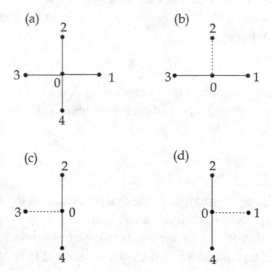

Figure 11.7 Finite difference node patterns.

   In applying the Poisson equation here, four types of node patterns are required, as shown in Figure 11.7. In each case, node 0 refers to the node we are currently evaluating. For a general interior node where all the neighbor nodes are in the same region, we apply the pattern (a), which results in the relation

$$4A_0 = A_1 + A_2 + A_3 + A_4 + f, \tag{11.17}$$

where $f = h^2 \mu J$ for nodes inside the conductor and 0 otherwise. For nodes on the symmetry plane, pattern (b) gives

$$4A_0 = A_1 + 2A_2 + A_3 + f.$$

For the left side of the iron sheet, we can use Equation 11.12 for pattern (c) with $\mu_a = 1$ and $\mu_b = 100$. For the right side of the sheet, we use Equation 11.12 for pattern (d) with $\mu_a = 100$ and $\mu_b = 1$.

For problems with a very large number of unknown nodes, it is not practical to solve the matrix equation using direct methods. Instead iterative methods must be used. A common method is to use the Successive Overrelaxation (SOR) algorithm. [10, 11] Let us define $A_{j,k}^n$ to be the value of the potential at the interior node at location (j, k) after n iterations. On the next iteration, we update the value of the potential according to the prescription

$$A_{j,k}^{n+1} = (1 - \alpha) A_{j,k}^n + \alpha A_{j,k}^*, \tag{11.18}$$

where α is called the overrelaxation parameter. For efficient convergence, we need $1 < \alpha < 2$. The optimal value for α is problem dependent, but the value $\alpha \sim 1.7$, which we use here, is typical. The quantity $A_{j,k}^*$ is the solution for A_{jk} from the appropriate nodal solution of Poisson's equation. For example, using Equation 11.17 for a general node, the SOR relation is

$$A_{j,k}^{n+1} = (1 - \alpha)A_{j,k}^n + \frac{\alpha}{4} \left[A_{j-1,k}^{n+1} + A_{j+1,k}^n + A_{j,k-1}^{n+1} + A_{j,k+1}^n + f_{j,k} \right].$$

Thus the updated value of the potential has two contributions. The first term is an adjustable fraction of its value on the previous iteration. The second term is a fraction of the Poisson equation residual at the node, calculated from the values of the potential at the neighbor nodes. Note that the calculation of the residual uses values for two nodes that have already been updated for a given iteration and values for two nodes from the previous iteration. The iterations continue until

$$\max \left| \frac{A_{j,k}^{n+1} - A_{j,k}^n}{A_{j,k}^n} \right| \leq \tau$$

over all the interior nodes.[1] The tolerance $\tau = 10^{-5}$ was used in this example. This criterion was satisfied after 7,511 iterations.

The magnetic field was calculated at the center of every square formed by four neighbor nodes, as shown in Figure 11.8.

$$B_x = \partial_y A_z = \frac{1}{2h} (-A_1 + A_2 + A_3 - A_4)$$

$$B_y = -\partial_x A_z = -\frac{1}{2h} (A_1 + A_2 - A_3 - A_4).$$

The results of the calculations for the magnetic field are shown in Figure 11.9.

---

[1] Or just the difference in values if the potential is 0.

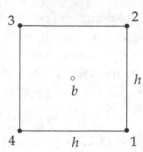

Figure 11.8 Magnetic field calculation.

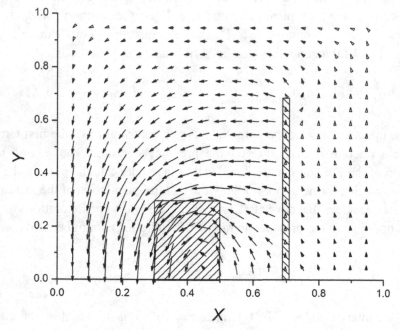

Figure 11.9 Magnetic field pattern for the example finite difference problem.

11.3 Finite element method

In the finite element method, the problem space is completely subdivided into a set of subregions called finite elements.[1, 12] The potential in each element is represented by an interpolation function that is defined in terms of the potential values at the nodes of the element. The Poisson equation and its boundary conditions can be formulated in terms of energy functionals. The minimization of this functional generates a set of algebraic equations that can be solved directly or through iterative techniques. The method is quite flexible since there is considerable freedom in choosing element shapes to match boundary and interface geometries.

For simplicity, we restrict our discussion here to two-dimensional problems. The nonlinear Poisson equation can be written in the form

$$\frac{\partial}{\partial x}\left(\gamma\frac{\partial A}{\partial x}\right) + \frac{\partial}{\partial y}\left(\gamma\frac{\partial A}{\partial y}\right) = -J,$$

where γ is the reluctivity and A and J only have nonvanishing components in the z direction. This differential equation can be expressed in terms of the energy functional [13]

$$\mathcal{F} = \iint\left[\int_0^B \gamma b \, db - JA\right] dx \, dy - \oint A\frac{\partial A}{\partial n} \, dl, \tag{11.19}$$

where b is the magnitude of the magnetic field.

$$b = \sqrt{\left(\frac{\partial A}{\partial x}\right)^2 + \left(\frac{\partial A}{\partial y}\right)^2}. \tag{11.20}$$

The line integral in Equation 11.19 vanishes since we require that the potential satisfy either Dirichlet or Neumann boundary conditions everywhere on the boundary. If the reluctivity is constant over an element, we can perform the integration over b to get the simplified energy functional

$$\mathcal{F} = \iint\left\{\frac{\gamma}{2}\left[\left(\frac{\partial A}{\partial x}\right)^2 + \left(\frac{\partial A}{\partial y}\right)^2\right] - JA\right\} dx \, dy. \tag{11.21}$$

The simplest two-dimensional finite element is a triangle, as shown in Figure 11.10. We assume the potential varies linearly inside the element.

$$A = c_1 + c_2 x + c_3 y. \tag{11.22}$$

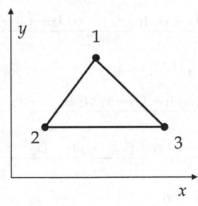

Figure 11.10 Triangular finite element.

If we write this expression for each of the three nodes, we have three equations that can be solved for the three unknown coefficients c_i in terms of the potentials and coordinates at the nodes. Substituting the result back into Equation 11.22, we find

$$A = A_1 \left[\frac{(x_2 y_3 - x_3 y_2) + (y_2 - y_3)x + (x_3 - x_2)y}{2S} \right]$$
$$+ A_2 \left[\frac{(x_3 y_1 - x_1 y_3) + (y_3 - y_1)x + (x_1 - x_3)y}{2S} \right] \tag{11.23}$$
$$+ A_3 \left[\frac{(x_1 y_2 - x_2 y_1) + (y_1 - y_2)x + (x_2 - x_1)y}{2S} \right],$$

where S is the area of the triangle.

$$S = \tfrac{1}{2}[(x_2 y_3 - x_3 y_2) + (y_2 - y_3)x_1 + (x_3 - x_2)y_1] \tag{11.24}$$

The coefficients of the node potentials in this equation are known as *shape functions*, ζ.[14] Thus we can also write the interpolation function for the potential as

$$A = \zeta_1 A_1 + \zeta_2 A_2 + \zeta_3 A_3. \tag{11.25}$$

The shape function ζ_1 has the properties that

$$\zeta_1(x_1, y_1) = 1$$
$$\zeta_1(x_2, y_2) = 0$$
$$\zeta_1(x_3, y_3) = 0$$

and similarly for ζ_2 and ζ_3.

In order to evaluate the simplified energy functional in Equation 11.21, we need the derivatives of A from Equation 11.23.

$$\frac{\partial A}{\partial x} = \frac{(y_2 - y_3)A_1 + (y_3 - y_1)A_2 + (y_1 - y_2)A_3}{2S}$$
$$\frac{\partial A}{\partial y} = \frac{(x_3 - x_2)A_1 + (x_1 - x_3)A_2 + (x_2 - x_1)A_3}{2S} \tag{11.26}$$

Substituting into Equation 11.21, we get

$$\mathcal{F} = \frac{\gamma}{2} \iint \left\{ \frac{[(y_2 - y_3)A_1 + (y_3 - y_1)A_2 + (y_1 - y_2)A_3]^2}{4S^2} \right.$$
$$\left. + \frac{[(x_3 - x_2)A_1 + (x_1 - x_3)A_2 + (x_2 - x_1)A_3]^2}{4S^2} \right\} dx\, dy$$
$$- \iint JA\, dx\, dy.$$

The solution of the field equations is equivalent to finding a function A that minimizes this energy functional.[15] The potentials at the three nodes may be considered to be the parameters of the functional $\mathcal{F}$ for a given element. Thus we require that

$$\frac{\partial \mathcal{F}}{\partial A_1} = \frac{\gamma}{2} \iint \left\{ \frac{2[(y_2 - y_3)A_1 + (y_3 - y_1)A_2 + (y_1 - y_2)A_3](y_2 - y_3)}{4S^2} \right.$$

$$\left. + \frac{2[(x_3 - x_2)A_1 + (x_1 - x_3)A_2 + (x_2 - x_1)A_3](x_3 - x_2)}{4S^2} \right\} dx\, dy$$

$$- \iint J \frac{\partial A}{\partial A_1} dx\, dy = 0$$

with analogous expressions for the derivatives with respect to A_2 and A_3. The integrand for the first integral is independent of x and y and the integrand for the second integral may be evaluated using Equation 11.23. Thus we have

$$\frac{\partial \mathcal{F}}{\partial A_1} = \frac{\gamma}{4S} \left[(y_2 - y_3)^2 A_1 + (y_3 - y_1)(y_2 - y_3)A_2 + (y_1 - y_2)(y_2 - y_3)A_3 \right]$$

$$+ \frac{\gamma}{4S} \left[(x_3 - x_2)^2 A_1 + (x_1 - x_3)(x_3 - x_2)A_2 + (x_2 - x_1)(x_3 - x_2)A_3 \right]$$

$$- \iint J \frac{[(x_2 y_3 - x_3 y_2) + (y_2 - y_3)x + (x_3 - x_2)y]}{2S} dx\, dy = 0$$

with analogous expressions for the derivatives with respect to A_2 and A_3. For elements containing current, the second integral can be evaluated by assuming that J is constant and that x and y are evaluated at the centroid of the triangle.

$$x_c = \frac{x_1 + x_2 + x_3}{3}$$

$$y_c = \frac{y_1 + y_2 + y_3}{3}.$$

In this case, the numerator in the last term is $2S/3$, so the integral has the value $JS/3$. Thus minimization of the functional over the triangular element leads to the matrix equation

$$\frac{\gamma}{4S} \begin{bmatrix} C_{11} & C_{12} & C_{13} \\ C_{21} & C_{22} & C_{23} \\ C_{31} & C_{32} & C_{33} \end{bmatrix} \begin{bmatrix} A_1 \\ A_2 \\ A_3 \end{bmatrix} = \frac{JS}{3} \begin{bmatrix} 1 \\ 1 \\ 1 \end{bmatrix}. \tag{11.27}$$

The coefficient matrix C is symmetric with six unique elements.

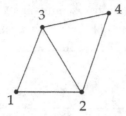

Figure 11.11 Two neighboring 2D elements.

$$C_{11} = (y_2 - y_3)^2 + (x_3 - x_2)^2$$
$$C_{12} = (y_3 - y_1)(y_2 - y_3) + (x_1 - x_3)(x_3 - x_2)$$
$$C_{13} = (y_1 - y_2)(y_2 - y_3) + (x_2 - x_1)(x_3 - x_2)$$
$$C_{22} = (y_3 - y_1)^2 + (x_1 - x_3)^2 \tag{11.28}$$
$$C_{23} = (y_3 - y_1)(y_1 - y_2) + (x_1 - x_3)(x_2 - x_1)$$
$$C_{33} = (y_1 - y_2)^2 + (x_2 - x_1)^2.$$

Each triangular element introduces an analogous set of equations. However, if N is the total number of elements, the number of unknown potentials is less than $3N$ because all of the elements share boundaries with neighbor triangles. For example, if we consider the two elements shown in Figure 11.11, the first element introduces three unknown potentials while the second element only adds one more. The resulting set of equations can be solved for the potentials using direct or iterative methods.

Setting up a realistic finite element problem involves a great deal of careful bookkeeping and computations.[13, 15] The problem space must be completely covered by the set of finite elements. The elements and nodes must be indexed and the association of each element with its corresponding nodes, current, and permeability must be clearly established. The boundary conditions must be imposed on the appropriate subset of the nodes. The coefficients for Equation 11.27 must be determined and an appropriate method used for solving the resulting system of equations. Additional iterative techniques must be applied if the problem contains saturable iron.

11.4 Integral equation method

Thus far we have discussed numerical methods for solving the Poisson differential equation directly and for solving the potentials by minimizing the energy functional for the magnetostatic field. Here we examine a third approach where the unknown potentials or sources of the field are expressed in terms of an integral equation. A major advantage of the integral equation method over methods based on the solution of a differential equation is that the mesh only needs to encompass the iron region (and possibly the conductor region if the current density is not uniform).[16, 17] The boundary condition at infinity follows naturally and does not have to be imposed at the edge of a mesh. An important disadvantage is that the resulting matrix equation is dense, so the solution time grows rapidly as the number of elements is increased. Also the flux density computed near the iron elements can be strongly affected by the discretization. There are many ways to formulate a solution of Poisson's equation using integral equations.[17, 18, 19] In addition, it is also possible to formulate procedures which combine differential and integral equation techniques.[20]

We describe here an integral equation approach that uses the magnetization of iron elements as the unknown function.[2] Recall that the magnetization is related to the magnetic field intensity by

$$\vec{M} = \frac{\vec{B}}{\mu_0} - \vec{H} = \chi(H)\,\vec{H}, \tag{11.29}$$

where χ is the susceptibility. The field intensity has contributions from both conductor currents and from the magnetization in the iron. The total field intensity is

$$\vec{H} = \vec{H}_c + \vec{H}_m,$$

so the magnetization is given by

$$\vec{M} = \chi\,(\vec{H}_c + \vec{H}_m).$$

The contribution H_c can be calculated for simple conductor configurations using the complex variable techniques given in Chapter 5 or directly from the Biot-Savart law

$$\vec{H}_c(\vec{r}) = \frac{1}{4\pi} \int \frac{\vec{J}(\vec{r'}) \times \vec{R}}{R^3}\, dV', \tag{11.30}$$

where $\vec{R} = \vec{r} - \vec{r'}$. Using Equation 3.32, the field due to the magnetization is

$$\vec{H}_m(\vec{r}) = -\frac{1}{4\pi} \nabla \int \vec{M}(\vec{r'}) \cdot \frac{\vec{R}}{R^3}\, dV'. \tag{11.31}$$

---

[2] This procedure was adopted by a group at the Rutherford High Energy Laboratory in the development of the GFUN program.

Break the iron region into N elements and assume the magnetization is constant over the area of each element. The magnetization at some element i will depend on the field at i due to the conductors and on the field at i due to the magnetization from all the elements. Thus we have

$$\vec{M}_i = \chi_i \left[\vec{H}_{ci} - \frac{1}{4\pi} \nabla_i \int \sum_{j=1}^{N} \vec{M}_j \cdot \frac{\vec{R}_{ij}}{R_{ij}^3} \, dV_j \right]. \tag{11.32}$$

We define the contribution to the field at element i due to the magnetization at element j in terms of the coupling constants

$$G_{ij} = -\frac{1}{4\pi} \nabla_i \int \hat{M}_j \cdot \frac{\vec{R}_{ij}}{R_{ij}^3} \, dV_j. \tag{11.33}$$

It is important to note that the components of G depends on the directions of the unit vectors used to define M, but do not depend on the magnitude of M. We can then rewrite Equation 11.32 as

$$\vec{M}_i = \chi_i \left[\vec{H}_{ci} + \sum_{j=1}^{N} G_{ij} \vec{M}_j \right].$$

Rearranging this equation, we have

$$\frac{\vec{M}_i}{\chi_i} - \sum_{j=1}^{N} G_{ij} \vec{M}_j = \vec{H}_{ci},$$

which can be written in the standard matrix equation form [17]

$$\sum_{j=1}^{N} \left(\frac{\delta_{ij}}{\chi_j} - G_{ij} \right) \vec{M}_j = \vec{H}_{ci}. \tag{11.34}$$

This is a set of algebraic equations for the N unknown magnetization elements. The number of unknowns is $2N$ for two-dimensional problems and $3N$ for three dimensions. The field components due to the conductors at each iron element can be calculated directly, so the right-hand side of Equation 11.34 is known. However, the problem is generally nonlinear because χ depends on the field due to the unknown magnetizations.

Returning to the definition of the coupling constant in Equation 11.33, we know from the vector identity B.2 that the gradient of the scalar product can be expanded as

$$\nabla_i\left(\hat{M}_j\cdot\frac{\vec{R}_{ij}}{R_{ij}^3}\right) = \hat{M}_j\times\left[\nabla_i\times\left(\frac{\vec{R}_{ij}}{R_{ij}^3}\right)\right] + \frac{\vec{R}_{ij}}{R_{ij}^3}\times(\nabla_i\times\hat{M}_j) + (\hat{M}_j\cdot\nabla_i)\left(\frac{\vec{R}_{ij}}{R_{ij}^3}\right)$$

$$+\left[\frac{\vec{R}_{ij}}{R_{ij}^3}\cdot\nabla_i\right]\hat{M}_j.$$

Two of the terms vanish because the derivative in field coordinates acts on the unit vector along the magnetization in source coordinates. Another term vanishes because of the curl operator acting on the linear vector R. Thus only the third term on the right-hand side remains. Writing this out in terms of components, we have

$$(\hat{M}_j\cdot\nabla_i)\left(\frac{\vec{R}_{ij}}{R_{ij}^3}\right)$$

$$= \left[\hat{M}_{jx}\partial_{ix} + \hat{M}_{jy}\partial_{iy} + \hat{M}_{jz}\partial_{iz}\right]\left[\frac{(x_i-x_j)\hat{x} + (y_i-y_j)\hat{y} + (z_i-z_j)\hat{z}}{\{(x_i-x_j)^2 + (y_i-y_j)^2 + (z_i-z_j)^2\}^{3/2}}\right]$$

$$= \frac{\hat{M}_{jx}}{R_{ij}^5}\left[R_{ij}^2\hat{x} - 3(x_i-x_j)\vec{R}\right] + \frac{\hat{M}_{jy}}{R_{ij}^5}\left[R_{ij}^2\hat{y} - 3(y_i-y_j)\vec{R}\right] + \frac{\hat{M}_{jz}}{R_{ij}^5}\left[R_{ij}^2\hat{z} - 3(z_i-z_j)\vec{R}\right].$$

After inserting this expression into Equation 11.33, we can identify the three-dimensional coupling constants

$$G_{ix,jx} = -\frac{1}{4\pi}\int\frac{R_{ij}^2 - 3(x_i-x_j)^2}{R_{ij}^5}\,dV_j$$

$$G_{ix,jy} = -\frac{1}{4\pi}\int\frac{-3(x_i-x_j)(y_i-y_j)}{R_{ij}^5}\,dV_j$$

and similarly for the other components.[21, 22] The G coupling matrix is symmetric. There are constraints on the sum of the diagonal elements.[21]

$$G_{ix,jx} + G_{iy,jy} + G_{iz,jz} = \begin{cases} 0 & \text{if} \quad i\neq j \\ -1 & \text{if} \quad i=j \end{cases}$$

In two dimensions, M_j is uniform along z, the field observation point has $z_i = 0$, and ∇_i and M_j only have x and y components. We find the two-dimensional coupling constants by integrating the three-dimensional couplings over z

$$G_{ij} = -\frac{1}{4\pi}\iint(\hat{M}_j\cdot\nabla_i)\left(\frac{\vec{R}_{ij}}{R_{ij}^3}\right)dz_j\,dS_j,$$

where dS_j is the two-dimensional area element. This can be written in the form

$$G_{ij} = -\frac{1}{4\pi} \int [\hat{M}_{jx}\, \mathbb{I}_1 + \hat{M}_{jy}\, \mathbb{I}_2]\, dS_j, \tag{11.35}$$

where

$$\mathbb{I}_1 = \int_{-\infty}^{\infty} \frac{R_{ij}^2\, \hat{x} - 3(x_i - x_j)[(x_i - x_j)\, \hat{x} + (y_i - y_j)\, \hat{y} - z_j\, \hat{z}]}{R_{ij}^5}\, dz_j$$

and

$$\mathbb{I}_2 = \int_{-\infty}^{\infty} \frac{R_{ij}^2\, \hat{y} - 3(y_i - y_j)[(x_i - x_j)\hat{x} + (y_i - y_j)\, \hat{y} - z_j\, \hat{z}]}{R_{ij}^5}\, dz_j.$$

Let r_{ij} be the distance between the observation point and the centroid of the iron element in the x-y plane. Then the integral $\mathbb{I}_1$ can be broken into the three simpler integrals[3]

$$\int_{-\infty}^{\infty} \frac{dz_j}{\{r_{ij}^2 + z_j^2\}^{3/2}} = \frac{2}{r_{ij}^2}$$

$$\int_{-\infty}^{\infty} \frac{dz_j}{\{r_{ij}^2 + z_j^2\}^{5/2}} = \frac{2}{3r_{ij}^4}$$

$$\int_{-\infty}^{\infty} \frac{z_j}{\{r_{ij}^2 + z_j^2\}^{5/2}}\, dz_j = 0$$

with the result that

$$\mathbb{I}_1 = \frac{2}{r_{ij}^2}\hat{x} - \frac{4(x_i - x_j)}{r_{ij}^4}\, \vec{r}_{ij}$$

$$\mathbb{I}_2 = \frac{2}{r_{ij}^2}\hat{y} - \frac{4(y_i - y_j)}{r_{ij}^4}\, \vec{r}_{ij}.$$

Inserting these results into Equation 11.35, we find the two-dimensional coupling constants are the dimensionless, geometric factors

---

[3] GR 2.271.5, 2.263.3, 2.271.7.

$$G_{ix,jx} = \frac{1}{2\pi} \int \frac{(x_i - x_j)^2 - (y_i - y_j)^2}{r_{ij}^4} dS_j$$

$$G_{ix,jy} = G_{iy,jx} = \frac{1}{2\pi} \int \frac{2(x_i - x_j)(y_i - y_j)}{r_{ij}^4} dS_j \qquad (11.36)$$

$$G_{iy,jy} = \frac{1}{2\pi} \int \frac{(y_i - y_j)^2 - (x_i - x_j)^2}{r_{ij}^4} dS_j.$$

There are also constraints on the sum of the two-dimensional diagonal elements.[21]

$$G_{ix,jx} + G_{iy,jy} = \begin{cases} 0 & \text{if} \quad i \neq j \\ -1 & \text{if} \quad i = j \end{cases}$$

Once we know the coupling constants G, we can solve Equation 11.34 to find the magnetization in each of the iron elements. Then the field at any position can be found from the sum of the fields due to all the current elements together with the sum of all the fields due to the iron magnetizations. In applications where saturation in the iron is important, the permeability of all the iron elements must be recomputed using the magnetizations and a μ–H table for the iron material. The process is then iterated until the maximum change in permeability in any element is less than some tolerance value.

Example 11.2: setting up the integral equations for a dipole configuration
We will illustrate the two-dimensional integral equation algorithm by considering a simple example of currents and iron blocks arranged in a dipole configuration. Once the current and iron magnetization has been determined in the first quadrant, the dipole symmetry constrains the geometry and polarity of the currents and magnetizations in the other quadrants, as illustrated in Figure 11.12. The currents have polarities $\{I, -I, -I, I\}$ in the four quadrants. If we let $(M_x^{(1)}, M_y^{(1)})$ refer to the magnetization of an iron element in the first quadrant, then the dipole symmetry requires that

$$M_y^{(2)} = M_y^{(3)} = M_y^{(4)} = M_y^{(1)}$$
$$M_x^{(2)} = M_x^{(4)} = -M_x^{(1)}$$
$$M_x^{(3)} = M_x^{(1)},$$

where the numeral superscripts refer to the quadrants. Making use of the dipole symmetry allows us to treat only the magnetization components in the first quadrant as unknowns. This can be important in problems with large numbers of iron elements since it reduces the size of the matrix equation by a factor of four.

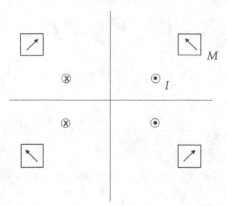

Figure 11.12 Dipole configuration of a current and iron magnetization.

Assume, for example, that there are two iron elements in the first quadrant. We can find the right-hand side of Equation 11.34 using the techniques in Chapters 4 and 5. We evaluate the field at the centroids of each of the iron elements. The magnetization has a constant magnitude and direction in each element. The coupling constants for the field due to the magnetizations can be found from Equation 5.72.

$$H^*(z_o) = \frac{M}{4\pi i} \oint \frac{dz^*}{z - z_o}.$$

We can determine the coupling constants G_{xx} and G_{yx} by evaluating H^* with $M = 1$ and the coupling constants G_{xy} and G_{yy} by evaluating H^* with $M = i$.

The field components due to the magnetization in each iron element can be found from

$$H^*_{mx} = G_{xx} M_x + G_{xy} M_y$$
$$H^*_{my} = G_{yx} M_x + G_{yy} M_y. \tag{11.37}$$

This gives the field anywhere outside the iron block. However, when the observation point is inside the block, the numerical procedure must ensure that H and M point in opposite directions, as they must inside a magnetic material. For the dipole configuration, we define the matrix coefficients C as sums over the coupling constants in the four quadrants. For example,

$$C_{ixjx} = G^{(1)}_{xx} - G^{(2)}_{xx} + G^{(3)}_{xx} - G^{(4)}_{xx},$$

where the minus signs take into account the reversal in the sign of M_x in the second and fourth quadrants. The other three coefficients are similarly defined. The matrix Equation 11.34 can be written for the case of two iron elements as

$$(s_1 - C_{1x1x})M_{1x} - C_{1x1y}M_{1y} - C_{1x2x}M_{2x} - C_{1x2y}M_{2y} = H_{c1x}$$
$$-C_{1y1x}M_{1x} + (s_1 - C_{1y1y})M_{1y} - C_{1y2x}M_{2x} - C_{1y2y}M_{2y} = H_{c1y}$$
$$-C_{2x1x}M_{1x} - C_{2x1y}M_{1y} + (s_2 - C_{2x2x})M_{2x} - C_{2x2y}M_{2y} = H_{c2x}$$
$$-C_{2y1x}M_{1x} - C_{2y1y}M_{1y} - C_{2y2x}M_{2x} + (s_2 - C_{2y2y})M_{2y} = H_{c2y},$$

where $s_i = 1/(\mu_{ri} - 1)$ and the numeral subscripts refer to the two iron elements in the first quadrant.

After solving the matrix equation, M is known for all the iron blocks. The contribution of the iron to the field at any location can be found using Equation 11.37, where G is now evaluated for the desired field point.

11.5 The POISSON code

We have shown results from the POISSON code[4] a number of times previously in this book. POISSON is one of the earliest examples of a finite element program. We give a brief description here of the method used in the code for solving the two-dimensional Poisson equation.[23, 24] The user defines the boundaries and properties of the physical regions in the problem, together with the boundary conditions at the borders of the problem space. The program then automatically sets up an irregular triangular mesh where every interior node is surrounded by six triangles. All boundaries and interfaces between regions lie on mesh lines. The current density is assumed to be constant in each triangle in a conductor region and the permeability is assumed to be constant in each triangle in an iron region.

The code does not solve the Poisson equation directly. Instead, the solution algorithm makes use of the Ampère law

$$\oint \vec{H} \cdot \vec{dl} = \int \vec{J} \cdot \hat{n} \, dS.$$

Allowing for saturation in the iron, this can be written as

$$\oint \gamma(B) \vec{B} \cdot \vec{dl} = \mu_0 \int \vec{J} \cdot \hat{n} \, dS,$$

where γ is the reluctivity. The vector potential is assumed to only have a z component and to satisfy $\nabla \cdot \vec{A} = 0$. Expressing B in terms of the vector potential, we get

---
[4] http://laacg.lanl.gov/laacg/services

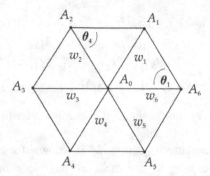

Figure 11.13 Relation among nodes in the POISSON algorithm.

$$\oint \gamma(B) \left[\frac{\partial A}{\partial y} \hat{x} - \frac{\partial A}{\partial x} \hat{y} \right] \cdot \vec{dl} = \mu_0 \int \vec{J} \cdot \hat{n} \, dS.$$

The contour around each mesh point follows a twelve-sided path through the interior of the six surrounding triangles. After a lengthy calculation,[24] a difference equation for the potential at node 0 can be derived in terms of the potentials at the six neighbor nodes as

$$A_0 = \frac{\displaystyle\sum_{i=1}^{6} A_i \, w_i + \frac{\mu_0}{3} \sum_{i=1}^{6} J_i \, S_i}{\displaystyle\sum_{i=1}^{6} w_i}.$$

In this equation, S_i is the area of the triangle. The w_i are coupling coefficients that involve the parameters of the triangles on adjacent sides of the line connecting A_0 to neighboring node i. Looking at the diagram in Figure 11.13,

$$w_1 = \tfrac{1}{2}(\gamma_1 \cot \theta_1 + \gamma_2 \cot \theta_4)$$

with similar expressions for the other five couplings.

The vector potential varies linearly inside any triangle. As a result, the magnetic field is constant over the area of the triangle. Values for the potential are updated using the successive over-relaxation algorithm. The new values of the field are then used to estimate new values for γ and for the couplings w. An under-relaxation algorithm is used to update the final values of the couplings for each iteration

$$w_i^{n+1} = (1 - \alpha)w_i^n + \alpha w_i^{new},$$

where the relaxation parameter satisfies $0 < \alpha < 1$.

As an illustration of using POISSON, we return to the simple problem discussed in Section 11.2. The input commands to define the problem are shown in

Table 11.1 *POISSON input commands for the example problem*

Example: rectangular conductor near iron sheet	
&reg kprob=0,	! Poisson or Pandira problem
icylin=0,	! rectangular coordinates
mode=-1,	! iron has fixed finite permeability
fixgam=0.01,	! reluctivity
dx=0.3,dy=0.3,	! mesh size intervals
nbslo=1,	! Neumann boundary condition on lower edge
nbsup=0,	! Dirichlet boundary condition on upper edge
nbslf=0,	! Dirichlet boundary condition on left edge
nbsrt=0 &	! Dirichlet boundary condition on right edge
&po x=0.0,y=0.0 &	! problem domain
&po x=100.,y=0.0 &	
&po x=100.0,y=100.0 &	
&po x=0.0,y=100.0 &	
&po x=0.0,y=0.0 &	
&reg mat=1,cur=19500. &	! conductor
&po x=30.0,y=0.0 &	
&po x=50.0,y=0. &	
&po x=50.0,y=30. &	
&po x=30.0,y=30.0 &	
&po x=30.0,y=0.0 &	
&reg mat=2 &	! iron sheet
&po x=70.0,y=0.0 &	
&po x=72.0,y=0.0 &	
&po x=72.,y=70.0 &	
&po x=70.,y=70.0 &	
&po x=70.0,y=0.0 &	

Table 11.1. The first REG command defines the problem domain and specifies the mesh size and the boundary conditions. The PO commands define points around the boundary of regions. The second region defines the conductor and specifies the current. The third region defines the iron sheet. POISSON sets up a triangular mesh using this information and then solves the Poisson equation using the SOR algorithm. For this example, the program used 112,896 mesh points, converged in 1,160 iterations, and had an average residual of 5×10^{-7}. The resulting field distribution, shown in Figure 11.14, agrees qualitatively well with the finite difference result in Figure 11.9. The shielding effects of the iron sheet are clearly apparent in the figure.

11.6 Inverse problems and optimization

We have previously defined the inverse problem as finding a current distribution that generates a specified magnetic field configuration. We discussed several problems of this type in Chapter 8. The solution of inverse problems is simplified

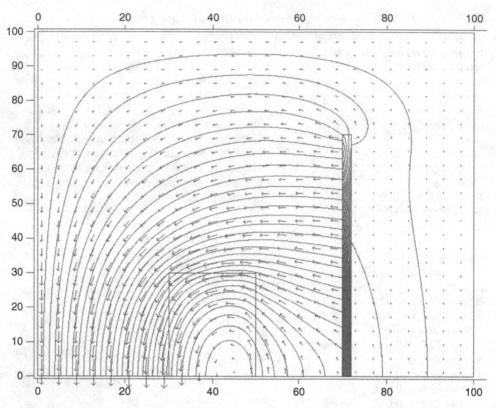

Figure 11.14 Field distribution for the example problem.

by constraining the coil geometry. For example, by specifying that the unknown currents lie on a cylindrical current sheet, it is possible to use Fourier-Bessel transforms to find the azimuthal and longitudinal current components that produce a specified target field inside a magnet aperture.[25] Another interesting approach used a numerical variational process to modify the contours of a uniform current density block conductor.[26] The target field was expressed in terms of a multipole expansion of the transverse field in the aperture. Higher-order multipoles were minimized by varying the geometry of the outer boundary of initial circular or elliptical current blocks.

A powerful technique for solving inverse problems is to make use of numerical optimization methods. Let us consider in more detail the numerical solution for two interesting inverse problems. As the first example, assume we have a solenoid channel with a constant axial field B_1 and that we need to design an interface region to a second solenoid channel with constant axial field B_2. Assume the interface has length L measured from the center of the last magnet in the first channel to the center of the first magnet of the second channel. Assume in addition that the transition region and second channel have to accept the full magnetic flux present in the first channel. Then the desired field profile in the interface region must satisfy the four constraints

$$B_z(0) = B_1$$
$$B_z(L) = B_2$$
$$\frac{dB_z}{dz}(0) = 0$$
$$\frac{dB_z}{dz}(L) = 0.$$

A model field profile that satisfies these constraints is

$$B_z(z) = \frac{B_1}{1 + c z^2 + d z^3}, \qquad (11.38)$$

where

$$c = \frac{3 (B_1 - B_2)}{B_2 L^2}$$

$$d = -\frac{2 (B_1 - B_2)}{B_2 L^3}.$$

If r_1 is the inner radius of the coils in the first channel, then the requirement for constant flux puts an additional constraint on the allowed inner radius of the downstream coils.

$$r(z) \geq r_1 \sqrt{\frac{B_1}{B_z(z)}}.$$

For example, let the coil C_1 be the last solenoid in a 10 T channel with a fixed inner radius of 10 cm and coil C_{14} be the first solenoid in a 2 T channel. Assume the transition region is 7 m long and contains 12 solenoids that are 45 cm long, separated by 5 cm, and have adjustable inner radius, radial thickness, and current density. The axial field for each solenoid uses Equation 7.46. The merit function f for the minimizer compares the desired value of the field at N locations z_i from Equation 11.38 with the calculated sum of the fields from all the coils, each with a set of parameters a_j.

$$f = \sum_{i=1}^{N} \left[\left(\sum_{j=1}^{14} B_z(z_i; a_j) \right) - B_z(z_i) \right]^2.$$

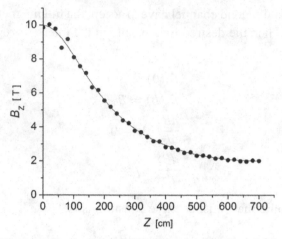

Figure 11.15 The optimized axial field (dots) and the desired field profile (line) in the transition region between two solenoid channels.

The calculation shown here used $N = 36$. Minimization of this function used methods that do not require calculation of the derivatives.[27] The initial minimization was done using a simplex algorithm. The most useful parameters to adjust were the current densities and the inner radii of the coils. The axial parameters are severely constrained here by the chosen geometry for the transition region. After a preliminary solution had been found, the Powell direction-set method was used for the final minimization. The optimized axial field is compared with the desired field profile in Figure 11.15.

As a second example of optimization, let us consider the design of the central section of a long dipole magnet with a circular cross-section. Assume that field quality in the dipole aperture is the matter of concern and that we want to minimize the strength of the first four allowed harmonics of the dipole field. We saw in Chapters 4 and 5 that the multipole coefficients depend on the limiting angles of annular conductor blocks. In order to eliminate four multipoles we will need to use at least three blocks. We choose here a conductor design with two radial layers, each of which has two conductor blocks, as shown in Figure 11.16. The contribution to the multipoles from an annular conductor block with constant current density was given in Equations 5.68–5.70. We again use a minimization algorithm, where the merit function is now given by

$$f = \sum_{i=1}^{4} w_i \left[\left(\sum_{j=1}^{16} b_n(n_i; a_j) \right) - \widetilde{b}_n(n_i) \right]^2 .$$

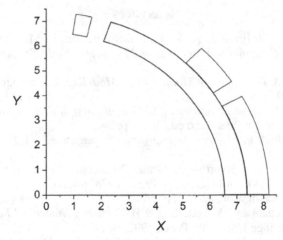

Figure 11.16 Conductor blocks in the first quadrant of a dipole magnet.

The index i sums over desired multipole orders, the index j sums over coils, and the normalized multipole ratio is defined as

$$b_n = \frac{B_n[T/m^{n-1}]\, r_{ref}^{n-1}[m^{n-1}]}{B_1[T]}.$$

The $\widetilde{b}_n$ factors are the desired values of the multipole ratios, which we take here as 0. The a_j are the set of parameters that describe conductor block j. The parameters of the coils in quadrants 2–4 are related to the parameters of the coils in quadrant 1 by the dipole symmetry. For this calculation the adjustable parameters are the end angles of the blocks nearest the midplane and the start and end angles of the second block in each layer. The start angle of the two blocks nearest the midplane are made as close to 0 as possible to maximize the dipole field. The w_i are weights that determine the importance of satisfying the constraint on multipole n_i. The reference radius used for the multipole calculations was 2/3 of the magnet aperture. After minimization, the allowed multipole ratios b_3, b_5, b_7, and b_9 have strengths $\sim 10^{-4}$.

In the design of actual magnets,[28] a need for high precision field quality may require that allowed multipoles higher than b_9 are also minimized. In addition, the conductor may have to be described in terms of individual turns of the cable separated by the appropriate insulation thickness, instead of the continuous conductor blocks used here. This introduces the additional constraint that there must be an integral number of turns in a conductor block. In addition, if the coils are surrounded by an iron shell, saturation effects, which cause the multipole strength to vary with the excitation current, may have to be taken into account.

References

[1] A. Frisiani, G. Molinari & A. Viviani, Introduction, in M. Chari & P. Silvester (eds.), *Finite Elements in Electrical and Magnetic Field Problems*, Wiley, 1980, p. 1–10.

[2] N. Gershenfeld, *The Nature of Mathematical Modeling*, Cambridge University Press, 1999, p. 86–91.

[3] K. Binns & P. Lawrenson, *Analysis and Computation of Electric and Magnetic Field Problems*, Pergamon Press, 2nd ed., 1973, p. 246.

[4] D. Jones, *Methods in Electromagnetic Wave Propagation*, vol. 1, Oxford University Press, 1987, p. 113.

[5] G. Smith, *Numerical Solution of Partial Differential Equations: Finite Difference Methods*, 2nd ed., Oxford University Press, 1978, p. 216.

[6] K. Binns & P. Lawrenson, *op. cit.*, p. 268.

[7] W. Press, S. Teukolsky, W. Vetterling & B. Flannery, *Numerical Recipes in Fortran*, 2nd ed., Cambridge University Press, 1992, p. 117.

[8] W. Milne, *Numerical Solution of Differential Equations*, Dover, 1970, p. 217.

[9] G. Parzen & K. Jellett, Computation of high field magnets, *Part. Acc.* 2:169, 1971.

[10] W. Press, et al., *op. cit.*, p. 857–860.

[11] K. Binns & P. Lawrenson, *op. cit.*, p. 260–265.

[12] N. Gershenfeld, *op. cit.*, p. 93–101.

[13] M. Chari, Finite element solution of magnetic and electric field problems in electrical machines and devices, in M. Chari & P. Silvester (eds.), *Finite Elements in Electrical and Magnetic Field Problems*, Wiley, 1980, p. 87–107.

[14] R. Gallagher, Shape functions, in M. Chari & P. Silvester (eds.), ibid., p. 49–67.

[15] P. Silvester & M. Chari, Finite element solution of saturable magnetic field problems, *IEEE Trans. on Power Apparatus and Systems* 89:1642, 1970.

[16] J. Simkin & C. Trowbridge, Three dimensional nonlinear electromagnetic field computations using scalar potentials, *IEE Proc.* 127:368, 1980.

[17] J. Simkin & C. Trowbridge, Magnetostatic fields computed using an integral equation derived from Green's theorems, *Proc. Compumag*, Rutherford Appleton Laboratory, Oxford, 1976, p. 5.

[18] C. Trowbridge, Applications of integral equation methods to the numerical solution of magnetostatic and eddy current problems, in M. Chari & P. Silvester (eds.), *op. cit.*, p. 191–213.

[19] C. Trowbridge, Progress in magnet design by computer, Proc. 4th Int. Conf. Magnet Technology, Brookhaven National Laboratory, 1972, p. 555.

[20] B. McDonald & A. Wexler, Mutually constrained partial differential and integral equation field formulations, in M. Chari & P. Silvester (eds.), *op. cit.*, p. 161–190.

[21] M. Newman, C. Trowbridge & L. Turner, *GFUN: an interactive program as an aid to magnet design*, Proc. 4th Int. Conf. Mag. Tech., Brookhaven National Laboratory, 1972, p. 617.

[22] F. Mingwu, S. Hanguang & W. Jingguo, Some experiences of using integral equation method to calculate magnetostatic fields, *IEEE Trans. Mag.* 21:2185, 1985.

[23] A. Winslow, Numerical solution of the quasilinear Poisson equation in a nonuniform triangle mesh, *J.Comp. Phys.* 1:149, 1966.

[24] J. Billen & L. Young, *Poisson-Superfish, Los Alamos National Laboratory report* LA-UR-96–1834, 2006, p. 577–596.

[25] R. Turner, A target field approach to optimal coil design, *J. Phys. D: Appl. Phys.* 19: L147, 1986.

[26] G. Morgan, Two dimensional, uniform current density, air core coil configurations for the production of specified magnetic fields, Proc. 1969 Part. Accel. Conf., Washington DC, p. 768.

[27] W. Press, et al., *op. cit.*, p. 402–413.

[28] E. Bleser, et al., Superconducting magnets for the CBA Project, *Nuc. Instr. Meth. Phys. Res. A* 235:435, 1985.

Appendices

A

Symbols and SI units [1, 2]

Symbol	Quantity	Unit	Dimension
I	current	A	$Q\,T^{-1}$
K	sheet current density	A/m	$Q\,T^{-1}\,L^{-1}$
J	volume current density	A/m^2	$Q\,T^{-1}\,L^{-2}$
B	magnetic flux density	T	$M\,T^{-1}\,Q^{-1}$
Φ_B	magnetic flux	Wb = T m^2	$M\,L^2\,T^{-1}\,Q^{-1}$
μ_0	permeability of free space	$=4\pi\,10^{-7}$ T m/A	$M\,L\,Q^{-2}$
M	magnetization	A/m	$Q\,T^{-1}\,L^{-1}$
H	magnetic intensity	A/m	$Q\,T^{-1}\,L^{-1}$
A	vector potential	Wb/m = T m	$M\,L\,T^{-1}\,Q^{-1}$
V_m	scalar potential	A	$Q\,T^{-1}$
L, M	self, mutual inductance	H = Wb/A	$M\,L^2\,Q^{-2}$
ρ	electric charge density	C/m^3	$Q\,L^{-3}$
V	potential difference	V	$M\,L^2\,T^{-2}\,Q^{-1}$
E	electric field intensity	V/m	$M\,L\,T^{-2}\,Q^{-1}$
σ	conductivity	$(\Omega\,m)^{-1}$	$T\,Q^2\,M^{-1}\,L^{-3}$
ε	permittivity	farad/m	$T^2\,Q^2\,M^{-1}\,L^{-3}$
D	electric flux density	coulomb m^2	$Q\,M^{-2}$
F	force	N = J/m	$M\,L\,T^{-2}$
W	stored energy	J = N m	$M\,L^2\,T^{-2}$
P	power	W = J/s	$M\,L^2\,T^{-3}$

References

[1] E. R. Cohen (ed.), *The Physics Quick Reference Guide*, American Institute of Physics, 1996, p. 37–47.

[2] D. Halliday & R. Resnick, *Physics for Students of Science and Engineering*, Wiley, 1962, appendix G.

B

Vector analysis

Vector analysis plays an essential role in describing the theory of magnetic phenomena.[1, 2] A vector V is a quantity that has both a magnitude and a direction. A scalar S is a quantity that only has an associated magnitude. Vector fields are functions that describe a physical quantity at every point in space.

The vector differential operator (*del*) is

$$\nabla = \frac{\partial}{\partial x}\,\hat{x} + \frac{\partial}{\partial y}\,\hat{y} + \frac{\partial}{\partial z}\,\hat{z}.$$

When ∇ is applied to a scalar function, it results in a vector known as the *gradient*.

$$\nabla S = \frac{\partial S}{\partial x}\,\hat{x} + \frac{\partial S}{\partial y}\,\hat{y} + \frac{\partial S}{\partial z}\,\hat{z}.$$

The gradient gives a measure of the rate of change of a vector. The dot product of ∇ with a vector forms a scalar known as the *divergence*.

$$\nabla \cdot \vec{V} = \frac{\partial V_x}{\partial x} + \frac{\partial V_y}{\partial y} + \frac{\partial V_z}{\partial z}.$$

Roughly speaking, the divergence gives a measure for the spreading out of a function away from a localized source. The *Laplacian* is an important operator that describes the second derivative of a scalar function and is given by

$$\nabla^2 S = \nabla \cdot \nabla S = \frac{\partial^2 S}{\partial x^2} + \frac{\partial^2 S}{\partial y^2} + \frac{\partial^2 S}{\partial z^2}.$$

It is also useful to define the Laplacian of a vector function, which is given in Cartesian coordinates as

$$\nabla^2 \vec{V} = \nabla^2 V_x\,\hat{x} + \nabla^2 V_y\,\hat{y} + \nabla^2 V_z\,\hat{z}.$$

The cross product of ∇ with a vector forms another vector known as the *curl*.

$$\nabla \times \vec{V} = \begin{vmatrix} \hat{x} & \hat{y} & \hat{z} \\ \partial_x & \partial_y & \partial_z \\ V_x & V_y & V_z \end{vmatrix}.$$

The curl gives a measure of the tendency of the vector to circulate around some source. According to *Helmholtz's theorem*,[3] a vector function that is bounded at infinity can be uniquely defined by specifying its divergence and its curl.

If we consider a volume of space V enclosed by a surface S, then we find that any changes in a vector W inside the volume must match the flux of W through the boundary surface. This is the basis for an important result known as Gauss's *divergence theorem*.[4]

$$\int \nabla \cdot \vec{W} \, dV = \int \vec{W} \cdot \hat{n} \, dS,$$

where n is the normal vector to the surface. If, on the other hand, we break up the surface S into a number of smaller areas and look at the net result of the circulation in all the subareas, we find that the circulations cancel in the interior of the region and only give a net result around the perimeter of S. The result is known as *Stokes's theorem*.[5]

$$\int (\nabla \times \vec{W}) \cdot \hat{n} \, dS = \oint \vec{W} \cdot \vec{dl}$$

Some other integral relations involving the gradient, divergence, and curl are less common, but still useful.[6]

$$\int \nabla P \, dV = \int P \, \hat{n} \, dS$$

$$\int \hat{n} \times \nabla P \, dS = \oint P \, \vec{dl}$$

$$\int \nabla \times \vec{W} \, dV = -\int \vec{W} \times \hat{n} \, dS,$$

where P is a scalar function and S is the surface that bounds the volume V.

The differential vector operators for cylindrical and spherical coordinate systems are given in Table B1.

Some important vector identities are

$$\vec{A} \times (\vec{B} \times \vec{C}) = \vec{B} \, (\vec{A} \cdot \vec{C}) - \vec{C} \, (\vec{A} \cdot \vec{B}) \tag{B.1}$$

Table B1 *Vector operators in cylindrical and spherical coordinates [6]*

	Cylindrical	Spherical
∇S	$\partial_\rho S\,\hat{\rho} + \dfrac{1}{\rho}\partial_\phi S\,\hat{\phi} + \partial_z S\,\hat{z}$	$\partial_r S\,\hat{r} + \dfrac{1}{r}\partial_\theta S\,\hat{\theta} + \dfrac{1}{r\sin\theta}\partial_\phi S\,\hat{\phi}$
$\nabla\cdot\vec{V}$	$\dfrac{1}{\rho}\partial_\rho(\rho\,V_\rho) + \dfrac{1}{\rho}\partial_\phi V_\phi + \partial_z V_z$	$\dfrac{1}{r^2}\,\partial_r(r^2 V_r)$
		$+\dfrac{1}{r\sin\theta}\left[\partial_\theta(V_\theta\sin\theta) + \partial_\phi V_\phi\right]$
$\nabla\times\vec{V}$	$\left(\dfrac{1}{\rho}\partial_\phi V_z - \partial_z V_\phi\right)\hat{\rho} + (\partial_z V_\rho - \partial_\rho V_z)\,\hat{\phi}$	$\dfrac{1}{r\sin\theta}\left[\partial_\theta(V_\phi\sin\theta) - \partial_\phi V_\theta\right]\hat{r}$
	$+\dfrac{1}{\rho}\left[\partial_\rho(\rho\,V_\phi) - \partial_\phi V_\rho\right]\hat{z}$	$+\dfrac{1}{r\sin\theta}\left[\partial_\phi V_r - \sin\theta\,\partial_r(r\,V_\phi)\right]\hat{\theta}$
		$+\dfrac{1}{r}\left[\partial_r(r\,V_\theta) - \partial_\theta V_r\right]\hat{\phi}$
$\nabla^2 S$	$\dfrac{1}{\rho}\partial_\rho(\rho\partial_\rho S) + \dfrac{1}{\rho^2}\partial_\phi^2 S + \partial_z^2 S$	$\dfrac{1}{r^2}\,\partial_r(r^2\partial_r S)$
		$+\dfrac{1}{r^2\sin\theta}\,\partial_\theta(\sin\theta\,\partial_\theta S)$
		$+\dfrac{1}{r^2\sin^2\theta}\,\partial_\phi^2 S$

$$\nabla(\vec{A}\cdot\vec{B}) = \vec{A}\times(\nabla\times\vec{B}) + \vec{B}\times(\nabla\times\vec{A}) + (\vec{A}\cdot\nabla)\,\vec{B} + (\vec{B}\cdot\nabla)\,\vec{A} \tag{B.2}$$

$$\nabla\cdot(S\,\vec{V}) = \vec{V}\cdot\nabla S + S\,\nabla\cdot\vec{V} \tag{B.3}$$

$$\nabla\cdot(\vec{A}\times\vec{B}) = \vec{B}\cdot(\nabla\times\vec{A}) - \vec{A}\cdot(\nabla\times\vec{B}) \tag{B.4}$$

$$\nabla\cdot(\nabla\times\vec{V}) = 0 \tag{B.5}$$

$$\nabla\times(S\,\vec{V}) = \nabla S\times\vec{V} + S\,\nabla\times\vec{V} \tag{B.6}$$

$$\nabla\times\nabla\times\vec{V} = \nabla(\nabla\cdot\vec{V}) - \nabla^2\vec{V} \tag{B.7}$$

$$\nabla\times\nabla S = 0 \tag{B.8}$$

$$\nabla\times(\vec{A}\times\vec{B}) = \vec{A}\,(\nabla\cdot\vec{B}) - \vec{B}\,(\nabla\cdot\vec{A}) + (\vec{B}\cdot\nabla)\,\vec{A} - (\vec{A}\cdot\nabla)\,\vec{B} \tag{B.9}$$

References

[1] G. Harnwell, *Principles of Electricity and Magnetism*, 2nd ed., McGraw-Hill, 1949, p. 636–649.

[2] J. Marion, *Classical Electromagnetic Radiation*, Academic Press, 1965, p. 447–456.

[3] L. Eyges, *The Classical Electromagetic Field*, Dover, 1980, p. 387.

[4] P. Lorrain & D. Corson, *Electromagnetic Fields and Waves*, 2nd ed., Freeman, 1970, p. 13–16.

[5] Ibid., p. 21–22.

[6] E. R. Cohen (ed.), *The Physics Quick Reference Guide*, American Institute of Physics, 1996, p. 162.

C

Bessel functions

Using the method of separation of variables for the Laplace equation in cylindrical coordinates gives rise to Bessel's equation.[1, 2]

$$\rho \frac{d}{d\rho} \left(\rho \frac{dR}{d\rho} \right) + (k^2 \rho^2 - n^2) \ R = 0.$$

In this equation, $R = R(\rho)$ and k and n are separation constants. The parameter n must be an integer to keep the azimuthal dependence of the solution single-valued, i.e., we must have

$$\Phi(\phi) = \Phi(\phi + 2\pi n).$$

Bessel's equation is a second order differential equation that has two independent classes of solution. One class involves Bessel functions of the first kind,[3] $R(\rho) = J_n(k\rho)$. The behavior of the first three Bessel functions J_n are shown as a function of $k\rho$ in Figure C1. All functions of this type are well-behaved at $\rho = 0$. They are oscillatory with a decreasing amplitude that approaches zero as $k\rho \to \infty$. The first root of the function $J_0(x)$ occurs at $x = 2.405$, where $x = k\rho$. The first root of $J_1(x)$ occurs at $x = 3.832$. The series expansion is

$$J_n(x) = \sum_{k=0}^{\infty} \frac{(-1)^k}{k! \ (n+k)!} \left(\frac{x}{2} \right)^{n+2k}.$$

The Bessel functions satisfy the recurrence relation

$$\frac{2n}{x} J_n(x) = J_{n-1}(x) + J_{n+1}(x),$$

while the derivatives satisfy the relation

$$2 J'_n(x) = J_{n-1}(x) - J_{n+1}(x).$$

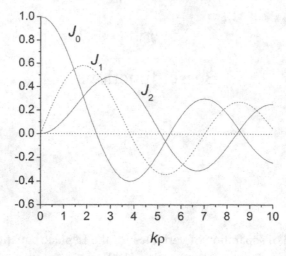

Figure C1 Bessel functions of the first kind for $n = 0, 1, 2$.

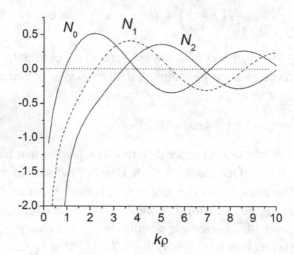

Figure C2 Bessel functions of the second kind for $n = 0, 1, 2$.

The derivative of J_0 is given by

$$\frac{dJ_0(x)}{dx} = -J_1(x).$$

The other class of solutions to Bessel's equation are the Bessel functions of the second kind,[4] $R(\rho) = N_n(k\rho)$. The behavior of the first three Bessel functions N_n are shown as a function of $k\rho$ in Figure C2. These solutions are also oscillatory with decreasing amplitude that approach zero as $k\rho \to \infty$. However, they diverge at $\rho = 0$, so they cannot be used in magnetostatics for any region that contains the origin. The $N_n(x)$ functions satisfy the same recurrence relations as $J_n(x)$. The derivative of N_0 is given by

$$N'_0(x) = -N_1(x).$$

If in applying the method of separation of variables for the Laplace equation in cylindrical coordinates, we require that the solution along z is oscillatory, then the separation parameter for the axial and radial terms must have the opposite sign from that used in deriving the Bessel differential equation. This leads to the radial equation

$$\rho \frac{d}{d\rho}\left(\rho \frac{dR}{d\rho}\right) - (k^2\rho^2 + n^2)\, R = 0.$$

The solutions of this equation are known as *modified Bessel functions*. This same equation can be produced by replacing k with $i\,k$ in the ordinary Bessel equation. One class of radial solutions involves the modified Bessel function $I_n(k\rho)$.[5] The behavior of the first three modified Bessel functions I_n are shown in Figure C3. All functions of this type are well behaved at $\rho = 0$. They are related to the ordinary Bessel functions by

$$I_\nu(x) = i^{-\nu} J_\nu(i\,x).$$

The series expansion is

$$I_n(x) = \sum_{k=0}^{\infty} \frac{1}{k!\,(n+k)!} \left(\frac{x}{2}\right)^{n+2k}$$

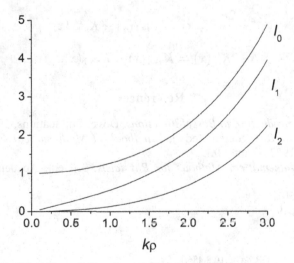

Figure C3 Modified Bessel functions $I_n(k\rho)$ for $n = 0, 1, 2$.

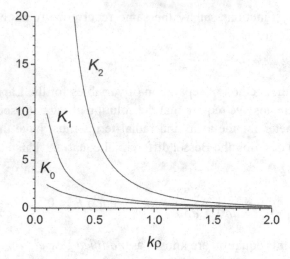

Figure C4 Modified Bessel functions $K_n(k\rho)$ for $n = 0, 1, 2$.

and it satisfies the recursion relations[1]

$$\frac{2n}{x}\, I_n(x) = I_{n-1}(x) - I_{n+1}(x)$$

$$2I'_n(x) = I_{n-1}(x) + I_{n+1}(x).$$

The other class of solutions for the modified Bessel's equation are the functions $K_n(k\rho)$. The behavior of the first three modified Bessel functions K_n are shown in Figure C4. These solutions diverge at $\rho = 0$, so they cannot be used in any region that contains the origin. The functions K_n satisfy the recursion relations[2]

$$-\frac{2n}{x}\, K_n(x) = K_{n-1}(x) - K_{n+1}(x)$$

$$-2K'_n(x) = K_{n-1}(x) + K_{n+1}(x).$$

References

[1] F. Bowman, *Introduction to Bessel Functions*, Dover Publications, 1958.
[2] M. Abramowitz & I. Stegun (eds.), *Handbook of Mathematical Functions*, Dover Publications, 1972, chapter 9.
[3] G. Arfken, *Mathematical Methods for Physicists*, 3rd ed., Academic Press, 1985, p. 573–584.
[4] Ibid., p. 596–601.
[5] Ibid., p. 610–619.

---

[1] GR 8.486.1, 8.486.2.   [2] GR 8.486.10, 8.486.11.

D

Legendre functions

Separation of variables for the Laplace equation in spherical coordinates gives the partial differential equation

$$\frac{1}{\sin\theta}\,\partial_\theta(\sin\theta\,\partial_\theta Y) + \frac{1}{\sin^2\theta}\,\partial_\phi^2 Y + l(l+1)\,Y = 0$$

for the angular dependence. The solution of this equation is given in terms of the spherical harmonic functions Y_{lm} [1, 2]

$$Y_{l\,m}(\theta,\phi) = \sqrt{\frac{(2l+1)}{4\pi}\,\frac{(l-m)!}{(l+m)!}}\,P_l^m(\cos\theta)\,e^{im\phi},$$

where l and m are integers and P_l^m is an associated Legendre function. Allowed values of m are all integers in the range $-l \le m \le l$. Values of the spherical harmonics for negative m are given by

$$Y_{l,-m} = (-1)^m\,Y_{l,m}^*,$$

where the asterisk denotes complex conjugation. The spherical harmonics for $l \le 2$ are given in Table D1.

The polar angle part $\Theta(\theta)$ of the solution to the Laplace equation has to satisfy the second order, ordinary differential equation

$$\frac{d}{dx}\left[(1-x^2)\frac{d\Theta}{dx}\right] + \left[l(l+1) - \frac{m^2}{1-x^2}\right]\Theta = 0$$

where $x = \cos\theta$. The solutions of this equation are called associated Legendre functions of the first and second kind,

$$\{P_l^m(x), Q_l^m(x)\}.$$

Table D1 *Spherical harmonics*

l	m	Y_{lm}
0	0	$\dfrac{1}{\sqrt{4\pi}}$
1	0	$\sqrt{\dfrac{3}{4\pi}}\cos\theta$
1	1	$-\sqrt{\dfrac{3}{8\pi}}\sin\theta\,e^{i\phi}$
2	0	$\sqrt{\dfrac{5}{16\pi}}\,(3\cos^2\theta-1)$
2	1	$-\sqrt{\dfrac{15}{8\pi}}\sin\theta\cos\theta\,e^{i\phi}$
2	2	$\sqrt{\dfrac{15}{32\pi}}\sin^2\theta\,e^{2i\phi}$

Only the functions of the first kind have convergent power series over the complete range $0 \le x \le 1$, so we choose

$$\Theta(\theta) = P_l^m(\cos\theta).$$

The associated Legendre functions can be calculated from

$$P_l^m(x) = \frac{(-1)^m}{2^l\,l!}\,(1-x^2)^{m/2}\,\frac{d^{l+m}}{dx^{l+m}}(x^2-1)^l.$$

Associated Legendre functions with negative m are related to functions with positive m by

$$P_l^{-m}(x) = (-1)^m\,\frac{(l-m)!}{(l+m)!}\,P_l^m(x)\,.$$

The associated Legendre functions for $l \le 3$ and $m > 0$ are given in Table D2.

In problems with azimuthal symmetry, we have $m = 0$. In this case, the associated Legendre functions reduce to the ordinary Legendre polynomials.

$$P_l^0(\cos\theta) = P_l(\cos\theta)$$

Table D2 *Associated Legendre functions*

l	m	P_l^m
1	1	$\sin \theta$
2	1	$3 \cos \theta \sin \theta$
2	2	$3 \sin^2 \theta$
3	1	$\frac{3}{2} (5 \cos^2 \theta - 1) \sin \theta$
3	2	$15 \cos \theta \sin^2 \theta$
3	3	$15 \sin^3 \theta$

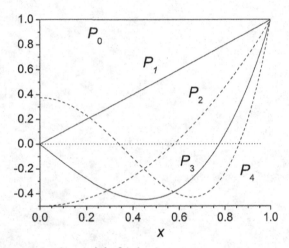

Figure D1 Legendre polynomials for $l \le 4$.

The Legendre polynomials form a complete set of orthogonal functions over the interval $-1 \le \cos \theta \le 1$. The behavior of the Legendre polynomials for $l \le 4$ are shown in Figure D1 as a function of x. Legendre polynomials satisfy the recurrence relation

$$(l + 1) P_{l+1}(x) = (2l + 1) x P_l(x) - l P_{l-1}(x)$$

and their derivatives satisfy the recurrence relation

$$(x^2 - 1) P_l'(x) = l x P_l(x) - l P_{l-1}(x).$$

References

[1] G. Arfken, *Mathematical Methods for Physicists*, 3rd ed., Academic Press, 1985, chapter 12.
[2] M. Abramowitz & I. Stegun (eds.), *Handbook of Mathematical Functions*, Dover Publications, 1972, chapter 8.

E

Complex variable analysis

We present here a brief summary without proofs of some of the important results from complex analysis that are relevant to the material covered in this book.[1, 2]

Complex variables

In a Cartesian coordinate system, the complex variable z is given by

$$z = x + i y,$$

where x is called the *real* part of z, y is called the *imaginary* part of z, and $i = \sqrt{-1}$. In polar coordinates, z can be written in the form

$$z = r \, e^{i\theta}$$
$$= r \, (\cos \theta + i \sin \theta),$$

where r is called the *modulus* of z and θ is the *argument* of z. The *De Moivre formula* is useful for evaluating powers of z.

$$(\cos \theta + i \sin \theta)^n = \cos n\theta + i \sin n\theta$$

The *complex conjugate* of a complex variable z is

$$z^* = x - i y.$$

The real and imaginary parts of a complex number can be written as

$$\mathbb{R}e(z) = \frac{z + z^*}{2}$$
$$\mathbb{I}m(z) = \frac{z - z^*}{2 i}.$$

Care is required in working with the complex counterparts of some real functions. An important function in magnetostatics is the complex logarithm function. This is defined as

$$
\begin{aligned}
w &= \ln z \\
&= \ln(r\, e^{i\theta}) \\
&= \ln r + i\,(\theta + 2\pi n),
\end{aligned}
$$

where $n = 0,\ \pm 1,\ \pm 2, \ldots$. This function has multiple *branches* of angular width 2π, depending on the value of n.[3] We customarily compute this function using the principal branch where $n = 0$ and where θ is in the range $-\pi < \theta \leq \pi$. In this case, the function changes discontinuously when crossing the negative x axis, which is called a *branch cut*.

Complex differentiation

The derivative of the complex function F is defined as [4]

$$
F'(z) = \lim_{\Delta z \to 0} \frac{F(z + \Delta z) - F(z)}{\Delta z},
$$

provided that the limit exists and is independent of the manner in which Δz approaches 0. If the derivative of F exists at all points throughout some planar region R, we say that the function is *analytic* in the region.[5] Examples of analytic functions include polynomials, exponentials, trigonometric, and hyperbolic functions. The real and imaginary parts of an analytic function are harmonic, i.e., they satisfy the Laplace equation.

Points where a function $F(z)$ is not analytic are called *singularities*. A singularity in $F(z)$ at a point z_0 is called a *pole* of order n if [6]

$$
\lim_{z \to z_0} (z - z_0)^n F(z)
$$

exists and is not 0.

An important property of analytic functions is that constraints exist between their real and imaginary parts.

Theorem E.1 (Cauchy-Riemann) [7]   (Necessity) *If a function $f(z) = u(x,y) + iv(x,y)$ is analytic in some domain D, then u and v have continuous first partial derivatives in D and satisfy the Cauchy-Riemann equations*

$$
\frac{\partial u}{\partial x} = \frac{\partial v}{\partial y}
$$

$$
\frac{\partial u}{\partial y} = -\frac{\partial v}{\partial x}.
$$

(Sufficiency) *If a function $f(z) = u(x,y) + iv(x,y)$ is defined in D, if* u *and* v *have continuous first partial derivatives in* D *and if the Cauchy-Riemann equations hold in* D, *then* f(z) *is analytic in* D.

For a region not including the origin, the Cauchy-Riemann equations can be written in polar coordinates as

$$\frac{\partial u}{\partial r} = \frac{1}{r}\frac{\partial v}{\partial \theta}$$

$$\frac{1}{r}\frac{\partial u}{\partial \theta} = -\frac{\partial v}{\partial r}.$$

Series

Theorem E.2 (power series) [8]   *Let $f(z)$ be analytic on a domain G and let z_o be an arbitrary point of G. Let $d = d(z_0)$ be the distance between z_o and the boundary of G. Then there exists a power series*

$$f(z) = \sum_{n=0}^{\infty} c_n(z - z_o)^n$$

that converges to f(z) on the disk $|z - z_o| < d$.

A power series can be differentiated or integrated term-by-term within its radius of convergence.

Theorem E.3 (Taylor series) [9]   *Let $f(z)$ be analytic and single-valued in an open region G. Let* a *be any point in G and let* C *be a circle with center at* a, *which together with its interior lies entirely in G. Then at every point* z *in* C, *the series*

$$f(a) + f'(a)(z - a) + \frac{f''(a)}{2!}(z - a)^2 + \cdots + \frac{f^{(n)}(a)}{n!}(z - a)^n + \cdots$$

converges to f(z).

In other words, $f(z)$ can be written as a Taylor series that converges in the region $|z - a| < R$, where R is the radius of convergence.

Theorem E.4 (Laurent series) [10]   *Let* f(z) *be analytic for the annular region*

$$G: R_1 < |z - z_o| < R_2$$

and let C *be any simple closed contour lying inside* G *and having z_o in its interior. Then for points* z *in* G, *the function* f(z) *may be expanded in the series*

$$f(z) = \sum_{k=-\infty}^{\infty} c_k(z - z_o)^k,$$

where

$$c_k = \frac{1}{2\pi i} \oint \frac{f(z)}{(z - z_0)^{k+1}} \, dz$$

and the integration is along the contour C.

The Laurent series is valid in a region surrounding, but not including, a singularity. Note that this series includes negative values of k. The coefficient c_{-1} has special significance and is known as the *residue*.

Complex integration

Theorem E.5 [11]   *If $f(z) = u(x, y) + iv(x, y)$ is continuous on a simple smooth arc from points* a *to* b, *then the integral exists and is given by*

$$\int f(z) \, dz = \int_a^b (u + iv) \, (dx + i \, dy).$$

Theorem E.6 (Cauchy integral theorem) [12]   *If* f(z) *is analytic in a simply connected domain* D, *then*

$$\oint f(z) \, dz = 0$$

on every simple closed path in D.

If instead of $f(z)$, we consider the contour integral of $f(z) / (z - z_o)$, then we have the following theorem.

Theorem E.7 (Cauchy's Integral Formula) [13]   *Let* f(z) *be analytic within and on a simple closed contour* C. *Then, if* z_o *is a point inside* C,

$$f(z_o) = \frac{1}{2\pi i} \oint \frac{f(z)}{(z - z_o)} \, dz.$$

This gives the value of $f(z_o)$ at the singularity z_o inside a region in terms of the contour integral around the boundary C.

Theorem E.8 [14]   *Let* f(z) *be analytic within and on a simple closed contour* C. *Then all derivatives of* $f(z_0)$ *exist at a point* z_0 *inside* C *and are given by*

$$f^{(n)}(z_0) = \frac{n!}{2\pi i} \oint \frac{f(z)}{(z - z_0)^{n+1}} \, dz.$$

Theorem E.9 (Residue theorem) [15]   *Let* f(z) *be analytic within and on a simple closed contour* C, *except for a finite number of isolated singularities inside* C. *Let* σ *be the sum of the residues at the singular points of* f(z) *that lie inside* C. *Then*

$$\frac{1}{2\pi i} \oint f(z) \, dz = \sigma.$$

In other words, the value of the contour integral is $2\pi i$ times the sum of the residues for the enclosed singularities. For a pole of order n, the residue can be found as [16]

$$a_{-1} = \frac{1}{(n-1)!} \lim_{z \to a} \frac{d^{n-1}}{dz^{n-1}} [(z-a)^n f(z)].$$

In the case of a simple pole ($n = 1$), the residue is given by

$$a_{-1} = \lim_{z \to a} (z - a) f(z).$$

Conformal mapping

We can define a function F that maps a complex variable z into a variable w in another two-dimensional space.

$$w = F(z).$$

Assume that two curves that cross at a point z_0 in the z space are separated by an angle θ. A mapping $w = F(z)$ is *conformal*, or angle preserving, if the mapped curves in the w space cross at the point $w_o = F(z_0)$ with the same angle θ.

Theorem E.10 [17]   *A mapping defined by an analytic function* F(z) *is conformal, except at points where the derivative* $F'(z)$ *is zero.*

Theorem E.11 (Riemann mapping theorem) [18]   *Let* D *be a simply connected domain with at least two boundary points. Then there exists a simple function* $w = F(z)$ *which maps* D *onto the unit disk* |w| < 1. *If we specify that a given point* z_0 *in* D *maps into the origin and a given direction at* z_0 *is mapped into a given direction at the origin, then the mapping is unique.*

Theorem E.12 (Schwarz-Christoffel transformation) [19]   *Let R be a polygon in the* w *plane having vertices at* w_1, w_2, ..., w_n *with corresponding interior angles* α_1, α_2, ..., α_n *respectively. Let the points* w_1, w_2, ..., w_n *map into the points* x_1, x_2, ..., x_n *on the real axis of the* z *plane. Then the transformation*

$$\frac{dw}{dz} = A(z - x_1)^{\alpha_1/\pi - 1}(z - x_2)^{\alpha_2/\pi - 1} \cdots (z - x_n)^{\alpha_n/\pi - 1},$$

where A *is a complex constant, maps the interior of the polygon in the* w *plane onto the upper half of the* z *plane and maps the boundary of the polygon onto the real axis of the* z *plane.*

References

[1] W. Smythe, *Static and Dynamic Electricity*, 2nd ed., McGraw-Hill, 1950, p. 72–94.
[2] W. Panofsky & M. Phillips, *Classical Electricity and Magnetism*, 2nd ed., Addison-Wesley, 1962, p. 61–72.
[3] J. Dettman, *Applied Complex Variables*, Dover, 1984, p. 58–60.
[4] K. Miller, *Introduction to Advanced Complex Calculus*, Dover, 1970, p. 48–52.
[5] M. Spiegel, *Complex Variables, Schaum's Outline Series*, McGraw-Hill, 1964, p. 63.
[6] Ibid., p. 67.
[7] W. Kaplan, *Advanced Calculus*, Addison-Wesley, 1952, p. 510–512.
[8] R. Silverman, *Introductory Complex Analysis*, Dover, 1972, p. 204–205.
[9] K. Miller, *op. cit.*, p. 114–115.
[10] J. Dettman, *op. cit.*, p. 163–164.
[11] Ibid., p. 81–82.
[12] W. Kaplan, *op. cit.*, p. 516.
[13] J. Dettman, *op. cit.*, p. 113.
[14] Ibid., p. 114.
[15] K. Miller, *op. cit.*, p. 132–133.
[16] M. Spiegel, *op. cit.*, p. 172–173.
[17] E. Kreyszig, *Advanced Engineering Mathematics*, Wiley, 1962, p. 719.
[18] J. Dettman, *op. cit.*, p. 256–259.
[19] M. Spiegel, *op. cit.*, p. 204, 218–219.

F

Complete elliptic integrals

An elliptic integral is an integral that can be written in the form [1]

$$\int R(x, \sqrt{f(x)})dx,$$

where R is a rational function and f is a third- or fourth-order polynomial in x. All integrals of this type can be written in terms of the three standard forms.

$$F(k, \theta') = \int_0^{\theta'} \frac{d\theta}{\sqrt{1 - k^2\sin^2\theta}}$$

$$E(k, \theta') = \int_0^{\theta'} \sqrt{1 - k^2\sin^2\theta} \, d\theta$$

$$\Pi(k, n, \theta') = \int_0^{\theta'} \frac{d\theta}{(1 + n \sin^2\theta)\sqrt{1 - k^2\sin^2\theta}}.$$

Each of these integrals depends on a parameter k called the *modulus* that satisfies $k^2 \leq 1$. The third type of integral also depends on a second parameter n called the *characteristic*.[2] When the upper limit of integration is

$$\theta' = \frac{\pi}{2},$$

these functions define the *complete elliptic integrals* of the first, second, and third[1] kinds.

---

[1] One should be aware that the complete elliptic integral of the third kind is sometimes defined with a negative sign before the factor n in the denominator.

$$K(k) = \int_0^{\pi/2} \frac{d\theta}{\sqrt{1 - k^2\sin^2\theta}}$$

$$E(k) = \int_0^{\pi/2} \sqrt{1 - k^2\sin^2\theta} \; d\theta$$

$$\prod(k, n) = \int_0^{\pi/2} \frac{d\theta}{(1 + n \sin^2\theta)\sqrt{1 - k^2\sin^2\theta}}$$

Moreover, these functions can alternatively be defined in polynomial form as

$$K(k) = \int_0^1 \frac{dx}{\sqrt{(1 - x^2)(1 - k^2x^2)}}$$

$$E(k) = \int_0^1 \frac{\sqrt{1 - k^2x^2}}{\sqrt{(1 - x^2)}} dx$$

$$\prod(k, n) = \int_0^1 \frac{dx}{(1 + nx^2)\sqrt{(1 - x^2)(1 - k^2x^2)}}.$$

It is important to emphasize that, despite the awkward nomenclature, the complete elliptic integrals are functions of k, and in the case of the third kind, also a function of n.

The complete elliptic integrals K and E can be expressed in terms of the infinite series

$$K(k) = \frac{\pi}{2}\left[1 + \left(\frac{1}{2}\right)^2 k^2 + \left(\frac{1 \cdot 3}{2 \cdot 4}\right)^2 k^4 + \cdots\right]$$

$$E(k) = \frac{\pi}{2}\left[1 - \left(\frac{1}{2}\right)^2 \frac{k^2}{1} - \left(\frac{1 \cdot 3}{2 \cdot 4}\right)^2 \frac{k^4}{3} + \cdots\right],$$

where $k^2 < 1$.[2] Efficient numerical algorithms have been developed to calculate the complete elliptic integrals.[3]

The dependences of the complete elliptic integrals of the first and second kinds are shown as a function of k in Figure F1. Both functions have the value $\pi/2$ for $k = 0$. The function $E(k)$ has the value 1 for $k = 1$, while $K(k)$ approaches ∞ as $k \to 1$. The behavior of the complete elliptic integral of the third kind for several values of n is shown as a function of k in Figure F2. The function $\prod(k, n)$ increases as k increases for all values of n. For a given value of k, the function increases as n becomes more negative.

If the vector potential is defined in terms of complete elliptic integrals, we need to take derivatives to find the magnetic field. In this case, we need to know

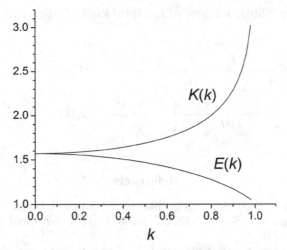

Figure F1 Dependence of the functions $K(k)$ and $E(k)$ on the modulus k.

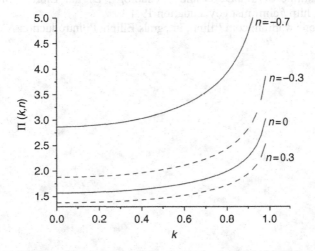

Figure F2 Behavior of the complete elliptic integral of the third kind.

the derivatives of the complete elliptic integrals with respect to their arguments.[2]

$$\frac{\partial K(k)}{\partial k} = \frac{E(k)}{k(1-k^2)} - \frac{K(k)}{k}$$

$$\frac{\partial E(k)}{\partial k} = \frac{E(k)}{k} - \frac{K(k)}{k}$$

[2] GR 8.123.2, GR 8.123.4.

For the complete elliptic integral of the third kind, the derivatives are given by [4, 5]

$$\frac{\partial \prod(k,n)}{\partial k} = \frac{k}{(1-k^2)(k^2-n)}\left[E(k) - (1-k^2)\prod(k,n)\right]$$

$$\frac{\partial \prod(k,n)}{\partial n} = \frac{1}{2(n-1)(k^2-n)}\left[E(k) + \frac{k^2-n}{n}K(k) - \frac{k^2-n^2}{n}\prod(k,n)\right].$$

References

[1] *CRC Standard Mathematical Tables*, 14th ed., Chemical Rubber Co., 1965, p. 282.

[2] M. Abramowitz & I. Stegun (eds.), *Handbook of Mathematical Functions*, Dover, 1965, p. 590–591.

[3] W. Press, S. Teukolsky, W. Vetterling & B. Flannery, *Numerical Recipes in Fortran*, 2nd ed., Cambridge University Press, 1992, p. 254–261.

[4] National Institute of Standards and Technology, Digital Library of Mathematical Functions, http://dlmf.nist.gov, equation 19.4.4.

[5] http://functions.wolfram.com/EllipticIntegrals/EllipticPi/introductions/CompleteElliptic Integrals/05

Index